Palgrave Studies in the History of Science and Technology

Designed to bridge the gap between the history of science and the history of technology, this series publishes the best new work by promising and accomplished authors in both areas. In particular, it offers historical perspectives on issues of current and ongoing concern, provides international and global perspectives on scientific issues, and encourages productive communication between historians and practicing scientists.

Suryakanthie Chetty

Science and Technology in South Africa, 1939–1946

The Mind at War

palgrave
macmillan

Suryakanthie Chetty
Department of Historical
and Heritage Studies
University of Pretoria
Pretoria, South Africa

ISSN 2730-972X ISSN 2730-9738 (electronic)
Palgrave Studies in the History of Science and Technology
ISBN 978-3-032-05083-0 ISBN 978-3-032-05084-7 (eBook)
https://doi.org/10.1007/978-3-032-05084-7

Cover illustration: Albert Haskin, 26 October 1940, File H-1: Mr Albert Haskin, KMG, Group 27, Box 179, Department of Defence Archives

This Palgrave Macmillan imprint is published by the registered company Springer Nature Switzerland AG
The registered company address is: Gewerbestrasse 11, 6330 Cham, Switzerland

Acknowledgements

I would like to acknowledge the efforts of the various archival staff who provided invaluable assistance. A special thank you to Tlangelani Baloyi (Archives Assistant) and Molatelo Pampa (Records Management Coordinator) at the University of Witwatersrand Central Records Office; and Steve de Agrela, Gerald Prinsloo, Anile Manya and Ferdinand Mukhawana at the Department of Defence Archives who have created such a collegial environment. And my thanks to the series editors, Roger D. Launius, James Rodger Fleming, Lucy Kidwell and the editorial staff at Palgrave Macmillan.

This work was the result of the support and kindness of Anne Samson who took on the role of my unofficial research assistant at The National Archives (Kew) and the British Library—and who encourages me every day; and Tilman Dedering who has been subjected to drafts on which he always provides constructive and very necessary feedback. His experience, expertise and graciousness are greatly appreciated.

And I am truly grateful to my friends and colleagues:

Anne and John Samson for support, compassion and friendship which has been unwavering over the years; Karen Jennings for empathy and generosity of spirit; Hannes Engelbrecht, my twin, who says that I've come home; Dawid Mouton who is my wingman; Shakira Govan for a lifetime of friendship.

My supportive colleagues at the Department of Historical and Heritage Studies: the incredibly inspiring and compassionate Karen Harris who has gone above and beyond as my HOD, Zimkhitha Tsotso, Nusrat Goolam Mahomed, C.R. Botha, Sias Conradie, Charlene Herselman, Ian Macqueen, Abraham Mlombo, Cornelis Muller, Nisa Paleker, Karina Sevenhuysen and Edwin Smith.

Melanda Blom, Anton and Salome Ehlers, John Connolly, Leschelle Morkel, Chet Fransch, Wessel Visser—and Melvyn Daniels who was kind and is missed greatly.

Finally, this is dedicated to Nicholas Southey who helps me find my way back when I am lost, and to my mother—who once held my hand and who holds it still.

Contents

List of Figures

CHAPTER 1

Introduction

"It is only men who are free, who create the inventions and the intellectual works which to us moderns make life worth while". This statement was uttered by arguably the most notable scientist of the twentieth century, Albert Einstein, in an address before an audience of ten thousand at the Royal Albert Hall in London. It was 1933 and the Nazis had come to power in Germany forcing Einstein to take temporary refuge in Britain before he left for the United States. In his speech, "Science and Civilization", shadowed by the authoritarianism that was gaining foothold in Europe and Asia, Einstein spoke of the way in which individual and political freedom had allowed for the flourishing of science and technology, without which people would be in perpetual servitude, leading an uninspired and monotonous existence. While he could not possibly foresee how his revolutionary work in theoretical physics would culminate in the fiery and definitive conclusion of the Second World War, Einstein alluded to the possibilities of change, of "a better world" arising from the "present upheavals".[1] Einstein's view of the role of science, technology and invention was one of hope yet his address neglected the complex ways in

[1] Albert Einstein, "Science and Civilisation", *Essays in Humanism* (New York, Open Road, 2015).

S. Chetty, *Science and Technology in South Africa, 1939-1946*, Palgrave Studies in the History of Science and Technology,
https://doi.org/10.1007/978-3-032-05084-7_1

which science and technology were enmeshed in war—and would become even more so in the conflict that was to come.

War highlighted the ambiguity associated with science and technology. The technology most associated with modern warfare was the aeroplane. The very first poem I committed to memory as a child was *High Flight* by John Gillespie Magee Jr. Magee can be seen as a symbol of the global world of the early twentieth century—born in China to an American mother and a British father, he first studied in Britain before hoping to further his studies at Yale University in the United States. With the outbreak of the Second World War, he enlisted in the Royal Canadian Air Force and was killed during flight training, aged just nineteen.[2] His poem—unmarred by the realities of combat—is of the exuberant freedom associated with flight, a freedom associated with a sense of grace. Published just a year later in 1940 and a depiction of the Spanish Civil War, *For Whom the Bell Tolls* portrays a very different vision of the aeroplane. Ernest Hemingway was a novelist and war correspondent who unabashedly favoured the Republican cause against Franco's fascism during the Spanish Civil War. In his novel of the conflict, Hemingway's protagonist describes the unnatural nature of aeroplanes and, particularly, bombers—they may resemble the apex predator, the shark, but "they move like no thing there has ever been. They move like mechanized doom".[3] Implicit in Hemingway's depiction of air power is the bombing raid carried out on the Spanish town of Guernica by German and Italian forces in April 1937—that would be immortalized in Pablo Picasso's painting based on photographs of and reportage from the devastated town.[4] The journalist partly responsible for bringing the story to the attention of the British was a South African, George Steer. This first publicization of a devastated European city, a civilian target, became what Richard Overy describes as a "lightning conductor for the accumulated anxieties of European populations".[5] In art and literature—and evident on a global scale—the aeroplane was emblematic of both technology

[2] "John Gillespie Magee Jr", *Poetry Foundation*, https://www.poetryfoundation.org/poets/john-gillespie-magee-jr, accessed 6 May 2025.

[3] Ernest Hemingway, *Hemingway on War* (London, Vintage, 2014), ppxxv, 147.

[4] Paloma Esteban Leal, "Guernica", Museo Reina Sofiahttps: https://www.museoreinasofia.es/en/collection/artwork/guernica, accessed 6 May 2025.

[5] Richard Overy, *The Bombing War: Europe 1939–1945* (Penguin Books, 2014), p33.

as symbol of progress and of its antithesis, death. Technology then was associated with the double-edged sword of modernity.

It was with the coming of revolution, and the overthrowing of the old order that modern Europe was birthed but so too were technological change and innovation in warfare. It was apparently a uniquely European phenomenon. During the last years of the twentieth century, historian David Landes could make the argument that while China may be accredited with numerous inventions that have changed the face of the world, these far-reaching changes only occurred as a result of European creativity and adaptation. Invention—and hence, technology—was a product of the modern, of change to the existing social order, of societies impelled by "progress" and "freedom", rather than "an older, effete reverence for authority".[6] From Landes' perspective—now anachronistic—invention was served by the liberal individualism associated with the capitalist West, and with a particular vision of masculinity based on the assertive, confident and free-thinking nature of an individual unbeholden to authority. Anachronistic or not, this perception was pervasive in the first half of the twentieth century—and beyond—and is exemplified by the relationship between science, technology, invention and war—an enduring association forged by the Second World War.[7] This relationship, however, was a shifting one and defined by the ideological divisions of the conflict as well as the exigencies of war.

The ambivalence, not just of writers, artists and poets, but of scientists to war can be traced to the interwar period. After the industrialized killing of "the war to end all wars", the utopian vision of science and technology was necessarily tempered. Broad trends can be detected in science and technology during this period—the first being the enormous developments in theoretical physics that would eventually culminate in the development of the atomic bomb. Yet scientific development and the concomitant development of sophisticated instruments increased the gap between the laboratory scientist and the general populace. Scientists too, ensconced in their "ivory tower" elevated scientific knowledge above the

[6] David Landes, *The Wealth and Poverty of Nations* (London, Abacus, 1998), p57. For a more critical and nuanced perspective, please refer to Ken Pomeranz, *The Great Divergence: China, Europe, and the Making of the Modern World Economy* (Princeton University Press, 2001).

[7] Refer to Tom Shachtman, *Laboratory Warriors: How Allied Science and Technology Tipped the Balance in World War II* (New York, Perennial, 2003).

mundanity of ideology. It was knowledge for its own sake, an expression of the human intellect yet, simultaneously, modern science in light of its origins, was a means of improving society.[8] Further, the practical application of science and its association with the military and industry were also evident, becoming an integral part of Franklin Roosevelt's military industrial complex. As evident in Bird and Sherwin's biography of Robert Oppenheimer, there was a tremendous expansion of industrial laboratories in the United States from just four in 1890 to almost a thousand by 1930, all of which were examples of "big science", demanding the use of ever more complex technology to carry out experiments, such as the cyclotron—and which would also increasingly focus on "military science".[9]

Science and technological development were also very much mired in national and international concerns. As David Edgerton points out, there were—and are—two dominant views of technology. "Techno-globalism" refers to the ways in which technology has taken on an almost utopian cast with the technologies of transportation (trains, steamships, aeroplanes) and communication (the telegraph, telephone) drawing the world together, creating what would become termed a "global village". In contrast to this utopian and perhaps "naïve" view, were the ways in which technology is associated with nationalism and especially with "national economic self-sufficiency"—which would later be especially evident in apartheid-era South Africa. It was ironic that it was these very technologies that fostered "techno-globalism" that were employed in a "new national despotism"—the technologies of empire and of war.[10] This dichotomy between the national and the global was exacerbated by the Second World War.

Among what would become the Allied countries of the war, the United States embodied both the theoretical and the practical application of science, its vast industrial capacity eventually making it indispensable in producing the technology invented in Britain, for instance, such as the

[8] Shachtman, *Laboratory Warriors*, p9.

[9] Kai Bird and Martin J. Sherwin, *American Prometheus: The Triumph and Tragedy of J. Robert Oppenheimer* (London, Atlantic Books, 2009), p86.

[10] David Edgerton, *The Shock of the Old: Technology and Global History Since 1900* (London, Profile Books, 2008), pp114–116.

cavity magnetron radar and penicillin.[11] In contrast to the more democratic and collaborative bent of the United States, class distinctions and elitism persisted in Britain, with South African reconstructive surgeon, Jack Penn, referring to the development of knowledge as being "clotted by tradition"—which nevertheless still permitted "originality" and innovation.[12] The totalitarian Soviet Union would demonstrate much in common with Germany after 1933.

In the totalitarian states of Germany and the USSR, science was inseparable from ideology. In Germany this ideology was racially based with the marginalization and expulsion of Jewish scientists that would adversely affect the lead that German science had made in fields such as theoretical physics during the interwar period. In the Soviet Union, as Josephson demonstrates, it was the ideology of class, and theoretical physics was marginalized for not being "a true description of nature", a nature that was predicated on "class conflict". Under both regimes, scientists were marginalized, forced to emigrate, interned and killed. Science lost its ostensibly autonomous nature—no longer based on the expertise of the scientist or the checks of the peer review process, it was instead made subordinate to the ideology of the state, the representatives of which scrutinized, limited and controlled scientific advances, their real-world applications and their "philosophical implications".[13]

As John Cornwell emphasizes, there is a distinction between German science and Nazi science and, in the period prior to 1933—and as reflected here—German scientific and medical developments and philosophical thought were influential in Europe and the United States, with revolutionary changes in physics, for instance. And, as evident in the chapter on aviation medicine, much of the field had its origins in the work of German specialists. With Adolf Hitler's ascent to power in 1933 however, German science was subject to the ideology of National Socialism and the arbitrary whim and will of its dictator. There was a superficial vision of "technocratic modernity"—the Volkswagen as "the people's car" in conjunction with the autobahns; the Berlin Stadium built for the Olympic Games to be held in 1936 that was to mark the union

[11] Shachtman, *Laboratory Warriors*, pp25, 129, 167–169.

[12] Jack Penn, *The Right to Look Human: An Autobiography* (Johannesburg, Hugh Keartland Publishers, 1976), p52.

[13] Paul R. Josephson, *Totalitarian Science and Technology* (New York, Humanity Books, 2005), pp73–74.

of the idealized Aryan body with the modern; the recording of sound and image. Simultaneously, increasing discriminatory measures against Jews and the expulsion of Jewish scientists—a precursor to the Holocaust—utilized medical imagery with the body of the *volk* polluted by the "parasitical" disease of Jewishness that needed to be "cauterized". For Hitler, this imagery of the Aryan body, also served as an obstacle to the implementation of modern weaponry, with Albert Speer describing the Fuhrer as "antimodern" based on his reluctance to have his soldiers use machine guns as it prevented "close combat". The same applied to the implementation of jet engines in fighter aircraft as Hitler considered the greater velocity an obstacle to "aerial combat". Early German efforts at an atomic bomb were also dismissed as "a spawn of Jewish pseudo-science".[14]

Yet Hitler's relationship to science was also ambivalent and inconsistent. Its effects though were far-reaching. Unswerving in his belief in racial hierarchy and ability, he claimed that the Soviet Union was incapable of "[invention]", of therefore being innovative and original; instead they could only "copy" the technology of others. This would in turn shape his military strategy, ultimately to Germany's detriment. His Japanese allies were seen as little different in terms of invention but they had the ability to "[improve]" on existing technology, "borrowing from left and right" in order to do so. In contrast, Hitler claimed to be "mad on technology", viewing the use of the new and the innovative as a means of keeping the enemy on the back foot. As a result, much needed resources would be poured into the development of untried rockets despite already existing and successful weaponry. It was a contrast to the ways in which the Allies viewed the use of invention. As Speer would state at the Nuremberg trials, "humanity" was sacrificed on the altar of "a technocracy".[15]

Japan's relationship with science and technology also occurred within an ideological climate of militarization. Japan's move to modernization was rapid in the nineteenth century and as James Bradley so eloquently points out, it had taken a mere two generations for the Japanese to go from being "fascinated" by the buttons on the uniforms of Commodore Perry's soldiers to firing upon the Russians with shells

[14] John Cornwell, *Hitler's Scientists: Science, War, and the Devil's Pact* (Penguin, 2003), pp4, 9–10, 24–25. In the final years of the War, the Germans would actually be the first to employ the jet engine in their fighter aircraft, the Messerschmitt Me 262.

[15] Cornwell, *Hitler's Scientists*, pp23, 34–35, 37.

manufactured in Japan and from Japanese-produced battleships in the Battle of Tsushima in May 1905. Their subsequent defeat of the Russians in the Russian-Japanese War would mark the first time a non-Western power had defeated a European power in modern warfare. During the interwar period, Japanese children were exposed to the twin poles of nationalism and war, with science as a school subject synonymous with military technology: "...searchlights, wireless communication, land mines and torpedoes, submarines, marine dirigibles, Shimose explosives, military carrier pigeons, heavy cannon, mortars, machine guns, the Arisaka cannon, and military sanitation".[16]

In Japan, science was a combination of research with a practical bent with the use of the language of science to buttress the power of the emperor and societal relationships in the form of "scientism". Its increasing militarization was evident from 1936 where it underwent what was described as a "fascist reorientation" and the formation of the Science Council in 1937 that would prioritize military research. Japan's imperial ambitions and the invasion of Manchuria also contextualized the establishment of Unit 731 and Japanese research into chemical and biological warfare using human subjects.[17] With Japan's expansion in the 1930s, efforts were made to employ the resources of occupied territories for Japan's development, underlying the formation of the "Greater East Asian Co-Prosperity Sphere" which could more accurately be understood as imperial exploitation. Other than in the sphere of biological warfare however, Japanese science and technology had not been harnessed fully to the needs of war, the prevailing ethos of militarization notwithstanding. This would change from mid-1942 with middling success, however it was only as late as September 1944 that the establishment of the Army-Navy Technology Enforcement Committee would make attempts to co-ordinate the scientific work of both the military and the civilian spheres. Even then, civilian scientists (especially those based in academia) demonstrated a lack of commitment to military goals.[18] It is an indication

[16] James Bradley, *Flyboys: A True Story of Courage* (New York, Back Bay Books, 2004), pp30, 35.

[17] Shachtman, *Laboratory Warriors*, pp21, 66–67.

[18] Walter E. Grunden, Yutaka Kawamura, Eduard Kolchinsky, Helmut Maier and Masakatsu Yamazaki, "Laying the Foundation for Wartime Research: A Comparative Overview of Science Mobilization in National Socialist Germany, Japan, and the

of the ways in which scientists may have harboured contrasting views of the role of science, even in autocratic regimes.

The harnessing of science—and scientists—to ideological and political imperatives was, however, not confined to totalitarian states. Even immediately prior to the outbreak of the War itself, both Britain and France had trusted that their economic and military strength might have deterred Hitler and so prevented an international conflict.[19] And their military strength was based on their armaments—so weaponry (as would be the case during the Cold War) was envisioned as a deterrent to war itself. Further, as evident in Shachtman's discussion, in Britain there were differing views of scientists regarding the use of science and technology as applied to war with those leaning towards socialism preferring the notion of "science in the service of society"—a utopian vision of scientific development and technology that could be used for the betterment of society and the relief of societal ills, as evident in the words of Albert Einstein with which this chapter began. For those leaning to the right, science and technology had to be adapted for military purposes as this was the only means to defeat the "Nazi machine". With the outbreak of war, there were distinctions made too between "offensive" and "defensive" technology with radar, for instance, perceived to be the latter and thus a more humanitarian use of technology in war. These distinctions, however, would turn out to be largely arbitrary. There was too the perception of the historic suspicion between scientists and the military with the latter's belief that civilian science subverted military autonomy. By 1940, however, the United States had formed the National Defense Research Committee, comprising both military and scientific experts, and marking a collaboration between the military and private and public institutions and laboratories. Scientists and military strategists had frequently to collaborate with each other and the military would come to "accept [an increasingly equal] partnership with science".[20]

Therefore, while the essays in this compilation focus on South Africa, it bears emphasizing that the Second World War was a global event. Based not just on conflict, it required the alliance and support of various states,

Soviet Union", *Osiris*, 2nd Series, Vol. 20, 79–106, "Politics and Science in Wartime: Comparative International Perspectives on the Kaiser Wilhelm Institute", pp95-98.

[19] Richard Overy, *1939: Countdown to War* (Penguin, 2010), p116.

[20] Shachtman, *Laboratory Warriors*, pp56, 115–116, 137, 203, 327.

colonies and dominions. As Lawrence James claims, Britain was only able to continue its war effort due to its empire, relying on the industrial production—most of which expanded as a response to the exigencies of war—the provision of combatants and auxiliaries and even the use of forced labour in some of the African colonies. Simultaneously, as a result of the demands made by the empire and of states, there existed both an exercise of—and a response to—the exertion of power. This would be manifest, for instance, in the rising nationalism evident in the colonies of Africa and Asia.[21] The war would also result in the shift in power that symbolized the decline of empire—and the initiation of the Cold War. This collection uses science and technology to explore the relationships, aspirations and tensions fostered by the War: between individuals and the state, between states within the ambit of the British Empire and within a changing global order.

As an Allied country, a former colony and a Dominion within the British Empire, South Africa was a relatively minor player in the War—although its leader Jan Smuts was a significant figure as international statesman, advocate of air power and confidante of Winston Churchill. As Saul Dubow has shown, from the beginnings of colonial settlement on the southern tip of Africa, science was imbued with ideology; it was a way of demonstrating the rhetoric of "civilization" associated with Empire, distinguishing those of British descent from the earlier Dutch settlers and from the indigenous people who were subject to a process of conquest and marginalization. Succeeding generations of settlers would view science as a way to assert an incipient sense of nationhood—within the confines of the British Empire. This would become even more evident after the formation of the Union of South Africa in 1910. The growth of scientific institutions, the expansion of industry and palaeoanthropological discoveries of early hominin fossils and the theory of continental drift gave new impetus to notions of modernization, progress and South African agency in the advancement of scientific knowledge.[22] Yet the idea of the nation was an exclusionary one and beset by societal tensions that were exacerbated by the Second World War, tensions that would culminate in Jan Smuts's electoral defeat in 1948 and the coming to power of

[21] Lawrence James, *Churchill and Empire: Portrait of an Imperialist* (London, Weidenfeld and Nicolson, 2021), pp227-232.

[22] Refer to Saul Dubow, *A Commonwealth of Knowledge: Science, Sensibility and White South Africa 1820–2000* (New York, Oxford University Press, 2007).

the apartheid government. It is within the tumult of the War, and the context of possibilities and of limitations, of a utopian vision of science and technology sitting uneasily with the reality of social inequality, that the essays that constitute this book are located.

Much of the historiography addressing the Second World War in South Africa tends to be military history with accounts given of the development and use of various weaponry. However, these essays have drawn upon a number of key texts addressing the relationship between the Second World War, science, technology and society. Bill Freund presents a developmental history of the South African state over the course of the twentieth century. His discussion of the Second World War addresses the industrialization of the state—an existing trend that was exacerbated by the War—as well as the ways in which science and, especially, technology were harnessed to the war with greater state control over industry and intervention in the economy. Of special note are the ways in which the drive for modernization conflicted with existing policies of segregation and a labour "colour bar".[23] In their comprehensive overview of the history of science and technology over the course of three centuries, Beinart and Dubow place South African scientific and technological development within the context of South Africa's relationship with Britain and the Commonwealth in the interwar period and during the Second World War, an era that would come to an end with the rise of the apartheid state in 1948.[24] In contrast to this overarching view, the edited collection *South Africa's 1940s: Worlds of Possibilities* explores the notion of the Second World War as a key moment in the social, political and economic history of South Africa that offered the potential for the creation of a new kind of state—than that which would come into being in 1948. Essays here focus on welfare, health and, of particular relevance to this work, the economic changes wrought by the War.[25]

On a global level, Shachtman's *Laboratory Warriors* re-envisions the Second World War through the lens of science and technology. With a focus on the key Allied and Axis powers (and thus necessarily addressing

[23] Refer to Bill Freund, *Twentieth-Century South Africa: A Developmental History* (Cambridge, Cambridge University Press, 2019).

[24] Refer to William Beinart and Saul Dubow, *The Scientific Imagination in South Africa: 1700 to the Present* (Cambridge, Cambridge University Press, 2021).

[25] Refer to *South Africa's 1940s: Worlds of Possibilities*, Saul Dubow and Alan Jeeves, eds. (Cape Town, Double Storey Books, 2005).

the northern hemisphere), he demonstrates the tense and often ambivalent relationship between science and war, discussing the different ways this played out in authoritarian and democratic states and concluding that it was the Allies' willingness to use their science expertise when compared to the reluctance of the Axis powers that was responsible for the Allied victory. In contrast to the broader scope adopted by Shachtman, Paul Kennedy's work focuses largely on the period from 1942. Drawing upon case studies related to five key features of the war effort, he seeks to understand the ways in which the Allies managed to halt what seemed like the remorseless Axis advance during the early years of the war and ultimately, "turn the tide". Here, warfare is almost reactive, as military and political leaders and strategists—as well as scientists and engineers—learned from early setbacks and failures, adapting to changing circumstances as they worked towards the aim of war: victory. Kennedy demonstrates that it is not just about the adoption of new technology or new strategies but whether these were efficacious and how this can be demonstrated during the key moments of the War.[26]

The work of David Edgerton has provided significant insight throughout this project. In *Warfare State* he challenges the existing historiography that touts appeasement and the rise of the welfare state in the interwar period by focusing on the enduring nature of the warfare state. This has provided a useful point of analysis for exploring the relationship of South Africa (a Dominion) in relation to Britain in terms of invention and the development of science and technology.[27] The notion of British "decline" that Edgerton challenges in *Warfare State* is revisited in his case study of the aeroplane where he addresses the military rather than civilian vision of aircraft.[28] While his work considers the British context, there are evident parallels between South Africa and Britain during the War that are indeed strengthened by the relationship between the two countries. Further, in *The Shock of the Old*, Edgerton applies his analysis to the perception and use of technology on a global scale. Here, Edgerton departs from—and, indeed, presents a critique of—conventional histories

[26] Paul Kennedy, *Engineers of Victory: The Problem Solvers Who Turned the Tide in the Second World War* (Penguin Books, 2014).

[27] David Edgerton, *Warfare State: Britain, 1920–1970* (Cambridge, Cambridge University Press, 2006).

[28] David Edgerton, *England and the Aeroplane: Militarism, Modernity and Machines* (Penguin, 2013).

of technology that focus on invention and innovation, portraying them as integral to "progress". Edgerton considers instead the *utilization* of technology, and continuities rather than disruptions in flaunting the new and the innovative. Older technologies remain ubiquitous, their use dependent on context and thus belying the image of a clear trajectory from the old to the new.[29] While this study does consider the use of invention in detail, it is influenced by Edgerton's view, addressing less the novelty of invention but the context of its production, its reception and its use—and how this can be used to illuminate the features of a society at war.

For Michael Adas, western science and technology cannot be divorced from an ideological context of dominance and difference. During the earliest stages of European exploration and the voyages of discovery, science and technology were part of a perception of "material culture" that included urbanization, architecture and literacy that were considered the defining features of "high civilizations"—a perception that remained intact until the last century. Even those "high [non-western] civilizations" fell considerably short of Enlightenment-era Europe with its sense of superior understanding that exceeded that of its Greek and Roman forebears as well as those of past and contemporary cultures around the globe. The Industrial Revolution was marked by technological change culminating in a drive for raw materials and new markets that would spur the imperial project. European technology was used to dominate the territories of Africa and Asia—guns to subdue resisting populations, steam trains and ships to enable exploration and conquest, medicine to circumvent the health risks associated with the "tropics" and to demonstrate visibly superior European knowledge. These, too, were integral to the "civilizing mission", the ideological imperative to remake the world in the image of Europe. Science and technology were also the means by which perceived racial differences, abilities and intellect could be measured and ordered. The carnage of the First World War, however, checked this unadulterated vision of progress and "civilization", provoking a crisis of faith. Yet, during the interwar period, the trust placed in science and technology and their relationship to modernization and progress remained, albeit less certain (except in the United States).[30] It is this more ambivalent

[29] David Edgerton, *The Shock of the Old: Technology and Global History Since 1900* (London, Profile Books, 2008).

[30] Refer to Michael Adas, *Machines as the Measure of Men: Science, Technology and Ideologies of Western Dominance* (Ithaca and London, Cornell University Press, 2014).

vision that is evident during the Second World War where science and technology were associated both with destruction and redemption.

And, as Joanna Bourke reminds us, there is the other face of war: the human cost. The civilian death toll of the Second World War greatly exceeded that of its predecessor and often exceeded that of combatants. Technology played no small part in this. From the industrialized extermination of the Holocaust to the aerial bombardment of cities and towns and the deployment of the most devastating and complex weapon of war—the atomic bomb—the utopian view of science and technology as the measure of human progress was offset by their use to wreak destruction and death. Medicine too, was not exempt, as evident in Japan's Unit 731 and the experiments carried out by Nazi scientists with a callous disregard for ethics and an unnerving dehumanization of their "subjects". Bourke's work then provides a human perspective of what can often be an alienating narrative of science, technology and medicine by considering those most affected by their use.[31]

The series of essays that constitute this book are interconnected to varying degrees. A key theme underlying the chapters is the human mind; in Chapter 2 which addresses the submission of ideas to the Union Defence Force, this is the inventive mind; in Chapter 7 with the focus on aviation medicine, there is the description of human will and agency and their ultimate subordination to technology and in Chapter 8 which considers psychiatry, it is the measurement of the mind that is the subject of intervention on the part of medical experts. The final chapter concludes with the concept of the "world mind" and the use of science to create a new post-war world order. Each chapter employs a case study to highlight the ways in which science and technology were interwoven with society and identity—national, racial and gendered—in wartime South Africa.

Chapter 2, "'Brain-waves to Beat Hitler': Science, Technology and Invention" uses the Ideas, Suggestions and Inventions Committee to explore the notion of invention and its relationship to science and technology. By considering the various ideas submitted to the Committee—set up by the Union Defence Force—it is possible to gain insight into the ways in which South Africans perceived both the concepts of invention and innovation as well as how they conceived of the war effort and their role in it. Inventions ranged from the credulous to the credible

[31] Refer to Joanna Bourke, *The Second World War: A People's History* (New York, Oxford University Press, 2001).

and the notion of invention itself is historicized and contextualized to demonstrate the ways in which the ideas submitted as well as prospective inventors both mirrored and challenged the inequalities and tensions in a country at war and one divided by racial distinction and ideology. The chapter also considers inventions in both Britain and other Dominions to address the ways in which South Africa was both influenced by and engaged with similar Allied wartime committees.

While invention may relate to individual ingenuity, the dominant narrative of South Africa at war was its mass industrialization—the production of weapons and war material. It is this narrative that comes to the fore in the third chapter, "'Strange and Wondrous Things': Military Technology on Show". Based on this dramatic expansion in secondary industry, this chapter addresses the ways in which in which South African military strength and its associations with science, technology and the modern were conveyed to the public. Considered here is the Rand Show in 1940, the nation-wide recruitment drives such as the Steel and Air Commandos, the War Train and the Liberty Cavalcades held in Johannesburg and Cape Town, respectively. The chapter concludes with the opening of the South African National War Museum in 1947. Throughout, it is evident that the display of military technology was associated with patriotism and nation-building, serving as an exercise in propaganda. It incorporated South Africans into the war effort and brought them into proximity with state and military power. Yet this demonstration of power also served to caution those elements in society who opposed and actively hampered that same war effort. Technology here, was both an instrument of inclusion and of domination.

In Chapter 4, "From Creation to Credit: Major Du Toit and the Invention of the Flail Tank", the focus is on the individual inventor and the arduous path of invention. The invention of the tank itself was in dispute during the First World War. Its adaptation as the flail tank of the Second World War once again raised the spectre of lack of due credit given to its inventor—as well as demonstrating the convergence between invention and utility. It was Union Defence Force Major A.S.J. du Toit who claimed to have invented the flail tank, an invaluable vehicle used to detonate mines, allowing for the advance of infantry. Yet there were rival claimants for the tank and the chapter looks at the inception of the idea as a response to real-world conditions on the battlefield, the path of invention and testing and the rival claims that were presented before the Royal Commission on Awards to Inventors in London. The case study of the flail tank also demonstrates the ways in which the UDF collaborated with their British counterparts in the development of the tank—it was a South African invention that could only be realized with the aid of British resources.

Chapter 5, "A Tale of Two Hospitals: Reconstructive Surgery and Race", on the other hand, considers the development of South African expertise in the field of reconstructive surgery. A comparative study of two hospitals is presented here: the Brenthurst Red Cross Military Hospital for Plastic Surgery established for white soldiers drawn from both South Africa and abroad, and No 110 Military Hospital Voortrekkerhoogte set up for the treatment of those classified as "Non-European"—soldiers who formed part of the auxiliary Non-European Army Services. The differences between the hospitals reflected not just the ambiguities of a segregationist state that had nevertheless taken on a more liberal bent during the War; they also highlight the ways in which medicine envisioned the white male body as the norm, an image that was highlighted by reconstructive surgery with its focus on the aesthetics of the physical body and the face. The chapter therefore addresses the treatment of soldiers at both hospitals with particular attention however paid to Voortrekkerhoogte: the types of injury incurred and the methods of treatment. While reconstructive surgery remained a relatively new field, the opportunities provided by the War allowed for invention and innovative methods of treatment, gaining Brenthurst some renown. Less acknowledged though was treatment at Voortrekkerhoogte where black bodies also formed the terrain on which expertise was built.

Reconstructive surgeon Jack Penn and the Brenthurst Hospital for Plastic Surgery are again the subject of the subsequent chapter, "Invention, Compensation and Dissemination at the Brenthurst Red Cross Military Hospital for Plastic Surgery". While Chapter 5 addressed the ways in which the bodies of black patients were the sites where reconstructive surgical techniques could be developed—even as the specialization followed the strictures of western medicine with the normalizing of the white male body—Chapter 6 considers the innovations and inventions originating at the Brenthurst Hospital, with a focus on the prosthetic eye and, in particular, the Brenthurst Splint. For Penn, the work done at the hospital was a means of asserting a southern African expertise in reconstructive surgery, a challenge to the dominance of specialists in Britain and the United States. The Splint was itself based on an identification of the shortcomings of existing devices used in the northern hemisphere, namely their invasive nature that left the patient vulnerable to infection. The Brenthurst Splint was non-invasive and able to be used by non-specialists—an important feature when injuries were incurred on the frontlines of battle. While the Splint symbolized national ingenuity, it was also a medical device and associated with the universalism of humanitarianism as envisaged by its creators, as well as a sense of international co-operation and the sharing of knowledge among the Allied reconstructive surgery pioneers. This however conflicted with the UDF's sense of proprietorship over the device and became especially thorny when the device was shared with the Soviet Union. The chapter demonstrates the competing interests of an inventor and what Penn would view as the dogmatic attitude of the military impeding innovation and the dissemination of knowledge. It reveals too, the complex and contradictory ways in which technology was shared between the Allies during the War.

Aviation medicine and the relationship between technology and the body are the focus of Chapter 7, "Man and Machine: Ernst Jokl, Aviation Medicine and 'a rational science of organism'". Beginning with the first flight at the dawn of the twentieth century in Kittyhawk, there was envisioned a harmony between the body of the pilot and the machine which he was piloting, with the machine an extension of the body. However, by the jet age, the developments in aviation technology with planes that could now fly at supersonic speeds as well as increasingly sophisticated equipment, meant that the agency of the pilot was curtailed, the human body now subordinated to the dictates of the machine. And drawing upon the work of David Edgerton that asserts the militaristic—and hence gendered—use of the aeroplane in the first half of the twentieth century, it was the male body that was the focus of aviation medicine. It is within this context that Ernst Jokl's work in aviation medicine can be situated. Drawing upon the pre-war studies of German scientists, Jokl's research addressed human responses to the conditions of modern aviation: extreme velocity and altitude. He demonstrated the imperfections of the senses and of a body wholly out of its natural environment. As a result, the cognition and physicality of the pilot was mediated by technology with pilots having to be trained to subordinate their instincts to the machine and its instrumentation. Flight, which was initially perceived as a legacy of knighthood exemplifying the actions of an individual warrior in anonymous, modern warfare, now served to limit the sense of the individual and, with it, an assertive masculinity.

The theme of aviation is continued in Chapter 8, "The Measured Mind: Aptitude Testing in the Union Defence Force", where the focus shifts from the physiological to the psychological requirements of flight. The chapter addresses the intelligence and aptitude testing carried out by the Aptitude Test Section. Due to the lack of resources and expertise in South Africa at the outbreak of the War, priority in intelligence testing was given to the most modern branch of the UDF, the South African Air Force. Considered briefly here are the origins of intelligence testing and its pre-war use in South Africa where there were attempts to distinguish and account for differences between black and white children and determine the aptitude of workers. As such, intelligence and aptitude testing were associated with the rational and scientific means of creating a modern country in terms of industry and, during the War, its military. Yet, despite the scientific methods of testing, assumptions of capability were underpinned by moral notions of courage and character stemming from the First World War. Aptitude testing was the method of "weeding out" those most likely to be overcome from fear—and who lacked the necessary moral fibre for combat. This chapter then discusses the work of the ATS in their attempt to determine the ideal aviator, looking at the testing process in terms of the development of tests, the methods used to determine measurable attributes such as intelligence as well as more intangible qualities such as "temperament" and "personality". It argues that, despite the scientific basis of testing, there existed an ambivalence related to what constituted an ideal temperament. This took on moral connotations that served to subordinate the needs of the individual to the well-being of the group.

The final chapter, "'The amoebic beginning of a world mind…': Sharing Knowledge in a Post-War World", considers the period immediately following the War and the sense of optimism in the envisioning of a utopian world. This new order was partly based on the sharing of scientific and technological knowledge with science perceived to be the panacea for societal ills: war, poverty, inequality and the rampant nationalism that had characterized the preceding conflict. Addressed here then are two conferences, the first of which is the Science and World Order Conference held in 1941 and designed to render concrete the idealistic principles of the Atlantic Charter. It was here that the British Commonwealth Scientific Collaboration Committee was established and they proposed a conference to be held five years later in London in 1946. While a symbol of collaboration, there were from the outset concerns about national security in terms of the sharing of scientific and technical information, especially that related to the military. The UDF nevertheless proposed a central repository where patents could be housed in Britain and to which all Commonwealth members would have access. In addition to national concerns and interests—as evident in the national scientific research organizations such as the South African CSIR—the Commonwealth Scientific Conference also reflected Britain's shifting relationship with the Dominions, acknowledging their roles as co-creators of scientific knowledge with national priorities that may have taken precedence over those stemming from their relationship with Britain. The Conference represented a brief moment of unity—although beset by existing divisions—before the polarization of the world during the Cold War and increasing South African isolation with the rise of the apartheid state in 1948.

Various archive repositories have been consulted for this project. These include the South African National Archives (Pretoria), the Department of Defence Archives (Centurion), The National Archives (Kew), the British Library, the Brenthurst Library and the University of the Witwatersrand Central Records Office.

CHAPTER 2

"Brain-Waves to Beat Hitler": Science, Technology and Invention

On 29 July 1915, the first meeting of the Admiralty Board of Invention and Research (BIR) was held in Victory House on Cockspur Street in London. The President was Lord Fisher who headed a committee consisting largely of Fellows of the Royal Society, some of the foremost scientists and engineers of the day.[1] The former First Sea Lord John Fisher was considered a "revolutionary" figure in his use of technology in the Royal Navy. Accredited with the early use of the submarine in the late nineteenth century as well as the development of the HMS Dreadnaught—and the subsequent class of dreadnaught battleships—that was steam-powered and heavily armed, Fisher was believed to be responsible for transforming a Victorian-era navy into a modern fighting force. Yet historians were initially unconvinced of Fisher's strategic thinking, perceiving his advocacy of new technology as a continuation of existing naval strategy rather than a revolutionary break. Later historians would

[1] David Rogers, *Top Secret: British Boffins in World War One* (Solihull, Helion and Company, 2013), p64.

"Brain-Waves to Beat Hitler": Title taken from *Rand Daily Mail*, 11 October 1940.

S. Chetty, *Science and Technology in South Africa, 1939-1946*, Palgrave Studies in the History of Science and Technology,
https://doi.org/10.1007/978-3-032-05084-7_2

argue however that Fisher's new technologies were part of his vision for an expanded role for the Royal Navy, dominating the waterways around Britain to safeguard it from invasion and making even more formidable the system of blockade to wage "economic warfare".[2]

In 1915 Fisher—who had resigned his position as First Sea Lord a few months earlier—found himself heading a committee of "boffins" tasked with "turning the scientific and inventive powers of the country" to the war effort. The aims of the BIR would resonate with the South Africans in the subsequent conflict: determining the utility of a submitted idea and encouraging "invention". Areas of focus for the BIR included airships, aeroplanes—and their seagoing counterparts—submarines, mines and artillery (and the protection against both) and "noxious gases in shell". All submissions were to be given careful consideration and "nothing [was to] be regarded as trash". Fisher was particularly taken with the use of invention and innovation in war as a way of thinking outside conventional parameters—while conventionality was a way of rising through the ranks during times of peace; it was a disregard of the conventional and the "careful" that would win wars: "What was wanted was Imagination – big Conceptions and quick Decisions".[3] The First World War, marked by industrialized killing was thus also perceived to be one based on innovation and invention, of the novel and the revolutionary. And it is this dichotomy that would also mark the succeeding conflict, both in Britain and in South Africa.

In an interview with a Reuters correspondent in 1943, Prime Minister Jan Smuts stated with optimistic conviction: "…Africa is the continent of the future. It teems with wealth and only awaits development". In the interview Smuts expounded on the possibilities and lessons learned from the War thus far—the most important of which was a change in the world order that would no longer be based on conquest and imperialism but on the self-determination of nations that would work together for the common good. He also believed in the potential of science to create his utopian world order: "Science has produced so many inventions that I believe it is quite possible that war will become physically impossible in

[2] David Morgan-Owen, "A Revolution in Naval Affairs? Technology, Strategy and British Naval Policy in the 'Fisher Era'", *Journal of Strategic Studies*, 944–965, 38:7, 2015, pp946–949.

[3] Rogers, *Top Secret*, pp65–67.

the future".[4] In a lecture at the annual meeting of the South African Association for the Advancement of Science three years earlier, just ten months after the outbreak of the War, scientist J.B. Robertson described war as an aberration, "...the result of an unscientific social order". Yet Robertson acknowledged that it was the use of science that rendered it ambiguous: a century of invention may have promoted growth in prosperity, but it had not done enough to alleviate hunger and disease. The solution was science in the service of society. For Robertson—as with Smuts—a new world order was in the making and scientists had the potential to alter the ways in which science had been used: "...the scientist, at once the product and the servant of a dying culture, holds in his hands the key to the making of a new culture". Again, like Smuts, Robertson believed that science would ultimately spell the end of war.[5] Yet, while science, technology and invention could potentially end war, they were also integral to how the Second World War was fought and won.

There are two narratives of science and technology in the Second World War in South Africa. The first is that of industrialization and rearmament.

The economy after the formation of the Union of South Africa in 1910 was concentrated in agriculture and especially in mining. Raymond Dumett highlights the importance of Africa's natural resources for the war effort, arguing that the conflict was itself based on the competition for resources. At a time when Britain and the United States were cut off from accessing raw material due to Japanese and German expansion, much focus was put on the African continent with the both the increasing exploitation of metals such as copper, chrome and platinum as well as the demand for "minor minerals" required for the new technology arising from the War—antimony, graphite, mica. South Africa had, since the late nineteenth century, been an exporter of gold, particularly to Britain and the extraction of gold increased during the War, increasing by 0.7 million tonnes of its pre-war levels, with much of this used by the British to pay the Americans for material purchased under the Lend Lease Act. This emphasis on gold production prompted an American delegation to attempt to persuade the Union government to focus more on the

[4] "Smuts Believes War Will End after 1944: Africa, Continent of the Future, Destined to Play Big Part", *Rand Daily Mail*, 28 August 1943.

[5] "Science Rightly Used Can Abolish War", *Rand Daily Mail*, 3 July 1940.

"strategic" metals needed for the War. One of these strategic metals was platinum. Initially, considered a "precious metal", it was used during the War for the production of essential items such as spark plugs, magnetos and fuses, as well as in the production of explosives. Among the Allies, the country was the third highest exporter of platinum with the bulk of its platinum going to Britain and the United States. Another essential South African export was coal derived from coalfields in the Orange Free State and Natal. The extraction of coal in South Africa was far more competitive due to the lower wages paid to workers and the Union had increased coal production by more than 10 million tonnes by the end of the War.[6]

A defining image of the military technology of the Second World War was the mushroom cloud over Hiroshima in August 1945. A year prior to the detonation of the atomic bombs over Japan, Britain, realizing the value of uranium, approached Jan Smuts to ascertain the extent of uranium deposits in South Africa. The radioactive element was also a by-product of the gold extraction process. As Asuelime shows, South African uranium because a key site of diplomacy between South Africa and Britain—and would continue to do so within the context of the Cold War.[7]

Yet, from the First World War—as South Africa was largely cut off from international markets—there was also some expansion in the manufacturing industry, a trend that would be evident in the subsequent war. While the Great Depression would have an adverse impact on growth, this was reversed in the 1930s, evident in the expansion in white labour but also relative growth in black labour as well.[8] The outbreak of the Second World War would exacerbate these trends, transforming the economy from one concentrated mainly on mining and its associated industries largely based in Johannesburg, to expansion in the manufacturing industry due to the prioritization of industry by the state, the stimulus created by war as well as the lack of imported manufactured goods due to the disruption wrought by the War. Yet the country did not

[6] Raymond Dumett, "Africa's Strategic Materials during the Second World War", *The Journal of African History*, Vol 26, No 4, 381–408, 1985, pp381–383, 385, 387.

[7] Lucky Asuelime, "Uranium politics of gatekeeping: Revisiting the British Government's Policy vis-à-vis South Africa, 1945–1951", *Historia*, Vol 58, No 1, 33–50, 2013, pp34–35.

[8] Bill Freund, *Twentieth-Century South Africa: A Developmental History* (Cambridge, Cambridge University Press, 2019), pp42–45.

possess the machinery to easily make the transition and initially, existing industries had to be adapted for the war effort. According to Nattrass, ammunition, for instance, was first produced by the South African Mint which alone possessed the machinery to do so.[9]

Wartime industry also required a quiescent labour force. In the 1920s the Smuts government was perceived to be hostile to the needs of (white) labour and the ensuing Pact government would prioritize white labour with job reservation and the skilling of the labour force. With Smuts's resumption of power at the outbreak of the War, increasing state intervention in a wartime economy meant the introduction of measures to subordinate labour to the demands of war. A series of War Measures were passed, permitting the state a more active hand in the economy and curbing labour unrest by both black and white workers.[10] Legislation such as War Measure 9 passed in 1942 served to curtail the power of white workers by outlawing strike action, while the economic power possessed by white skilled workers was significantly reduced through the policy of dilution that permitted black men and white women to take on work in essential industries such as munitions. The War created the conditions for the state to address industrial development—and thus the greater skilling of black labour—over the privilege enjoyed by white workers in the previous decades.[11]

H.J. van Eck and H.J. van der Bijl drove South Africa's industrialization during the War with the former the chair of the Industrial Development Corporation and the latter its manager. In 1940 the Corporation was tasked with conducting a "thorough survey" of South African industry so as to determine the development required. Significantly, proposed industrialization was not simply to foster wartime industry but to consider long-term industrial development.[12] Both men would also be involved in the production of iron and steel, contributing to the sense of South African wartime self-sufficiency. Van der Bijl in particular, was

[9] Nicoli Nattrass, "Economic Growth and Transformation in the 1940s", Saul Dubow and Alan Jeeves, eds., *South Africa's 1940s: Worlds of Possibilities* (Cape Town, Double Storey Books, 2005), p24.

[10] T.R.H. Davenport, *South Africa: A Modern History* (Johannesburg, Macmillan South Africa, 1978), pp241–242.

[11] Nattrass, "Economic Growth and Transformation in the 1940s", pp28–31.

[12] "Industrial Board Meets", *Rand Daily Mail*, 18 October 1940.

the Director of War Supplies and fostered the supplies of war material in conjunction with New Zealand, Australia and India: "army boots and blankets, tyres, munitions, armoured cars, precision instruments, and explosives".[13] In late 1940, he led a tour of the factories engaged in war production in Johannesburg that included the Governor-General of South Africa, Sir Patrick Duncan. The tour was a demonstration of South African industrial might—the massive production of armaments and military equipment that was vital to the war. Evident too was the role of the mining industry with factories and workshops churning out bombs and shells while simultaneously maintaining the high rate of gold production needed. South African steel was used to make armoured cars and helmets, wooden boxes built to hold bullets and more precision instruments such as range finders and gun sights were also produced, the latter standing "in rows on shelves". Significantly, war production was associated with nationalism and the country's economic strength where both the "raw material" viz. iron and steel, as well as the finished products were South African. Van der Bijl's tour was a demonstration of the country's "great industrial potential" that, while initiated by the War, would continue in its aftermath.[14]

Van der Waag argues that the outbreak of the Second World War in September 1939, meant a dramatic overhauling of the Union Defence Force (UDF) in South Africa and the rapid move by industry to serve the needs of war. Recruitment was increased and efforts were made to rectify the paucity of arms and other military accoutrements. The War also created the opportunity for the formation of the Seaward Defence Force and South Africa's eventual independence from the Royal Navy in commanding its own waters. Whalers and fishing trawlers were adapted for military use and Britain provided guns for coastal defence. Despite Jan Smuts's touting of the possibilities of air power from the 1920s as well as South African Minister of Defence Oswald Pirow's own passion for aviation, the South African Air Force (SAAF) was "quite unprepared" with aircraft considered "obsolete".[15] Yet, as Mouton shows, just six months prior to the War, Pirow had lauded the dramatic increase in the number

[13] Saul Dubow, *A Commonwealth of Knowledge: Science, Sensibility, and White South Africa, 1820–2000* (Oxford, Oxford University Press, 2007), p237.

[14] "His Excellency Sees Great Arms Drive", *Rand Daily Mail*, 18 October 1940.

[15] Ian van der Waag, *A Military History of Modern South Africa* (Johannesburg and Cape Town, Jonathan Ball Publishers, 2015), pp185–186, 188.

of pilots being trained and those in the reserves while also alluding to the challenges faced by South Africa in obtaining "modern" weaponry.[16] Upon the outbreak of war however, the expansion of the SAAF was dramatic, with an increased number of flying schools with more opportunities to train South African as well as other Allied pilots. In addition, while Britain was not able to meet the demand for aircraft in South Africa, Smuts would attempt to get aircraft from the United States.[17]

Yet the UDF was not averse to demonstrating its capability. Military exhibitions at the Rand Show and recruitment tours undertaken around the Union[18] introduced the technology of war to the ordinary South African, serving as a point of convergence between the grand narrative of wartime industrialization and civilian support of the war effort. Not unrelated to this is the second narrative: the story of the individual, of invention and of innovation.

In *Warfare State*, David Edgerton makes the compelling argument that the dominant perception of Britain during the Second World War of "eccentric boffins", an essentially pacifist people fighting a defensive war, should be reassessed by taking account of Britain's pre-war industrialization and militarism. Britain had continued rearmament during the interwar period with mass production, and political leaders were all too aware of the potential threat to peace presented by the authoritarian and nationalist regimes. The welfare state existed—but it did so alongside the warfare state.[19] In South Africa, on the other hand, as evident in the preceding discussion, the historiography is largely concerned with industrialization, development and the control of labour—the last integral to the economy of both the segregation and the apartheid state and largely underlying the policies of racial discrimination.

However, this chapter takes particular cognizance of the view of Arthur Marwick where invention is an integral part of the war effort. Invention is democratic, "open to the most ignorant among us", even if it is often those with scientific or engineering expertise who are most likely to make significant contributions. However, its ostensibly democratic

[16] F.A. Mouton, *The Opportunist: The Political Life of Oswald Pirow, 1915–1959* (Hatfield, Protea Book House, 2020), p129.

[17] Van der Waag, *A Military History of Modern South Africa*, pp187, 189.

[18] This is discussed in the following chapter.

[19] Refer to David Edgerton, *Warfare State: Britain, 1920–1970* (Cambridge, Cambridge University Press, 2006).

nature means that inventions can by their nature be unconventional and unexpected yet nevertheless make some impact. And war is the ideal environment to advance invention. The technological requirements of war lead to greater resources being allocated to applied science and to invention. Existing technology may be improved, and new technologies may arise.[20]

Yet this apparently democratic nature of invention is also offset by the ideological context that has been an integral part of western knowledge systems from the Enlightenment (and earlier). In *Machines as the Measure of Man*, Michael Adas describes an encounter between explorer William Smith, his acculturated black slave and the indigenous people of Guinea. The last were overcome by superstitious dread when confronted with Smith's surveying instruments, so much so that Smith felt that they were utterly incapable of understanding their construction and use, should these be explained to them. The slave, on the other hand, had been effectively trained in their use but he, too, could not work independently without white supervision and was even less capable of creating (or inventing) technology. Science and technology—and their application through invention—were therefore hallmarks of those of European descent, symbols of their "development", a narrative marked by racial distinction and hierarchy.[21] In South Africa, this distinction and exclusion was marked by both ideology and its concrete reinforcement through unequal access to education. The ideas and inventions submitted by South Africans then, mirrored the assumptions and inequities in South African society—even as they occasionally challenged them.

Beinart and Dubow state in *The Scientific Imagination in South Africa*, that one of the definitions of the "scientific imagination" relates to "innovation". This is not accorded sufficient import however as their preferred lens is the ways in which science is an exercise of power and has been so throughout South Africa's history, from its role during the expansion of the British Empire to its utilization by the apartheid state and now, its role in the contemporary period.[22] It is certainly evident that science is

[20] Arthur Marwick, *The Deluge: British Society and the First World War* (Basingstoke, Palgrave Macmillan, 1991), p267.

[21] Michael Adas, *Machines as the Measure of Men: Science, Technology and Ideologies of Western Dominance* (Ithaca and London, Cornell University Press, 2014), pp2–3.

[22] William Beinart and Saul Dubow, *The Scientific Imagination in South Africa: 1700 to the Present* (Cambridge, Cambridge University Press, 2021), p3.

inextricably linked to power and ideology—as these chapters indicate—and war is a graphic illustration of this link. Yet there exists, too, the ways in which individuals engage with science and technology, even if this is mediated by relationships of power as embodied in the state and the military. In a segregated country with factions including Afrikaner nationalists and right-wing fascists, support for the war effort in South Africa was by no means certain or even uniform. Even if undeveloped, the submissions and the assessment of inventions submitted to the UDF offer insight into civilian perceptions of the war effort in relation to science, technology and the notion of innovation.

This chapter therefore looks at the establishment and work of the Ideas, Suggestions and Inventions Committee and considers too some of the inventions submitted by the public. To do this, it draws upon the records of the Committee held at the Department of Defence Archives as well as contemporaneous texts and secondary material to contextualize the inventions proposed.

The Ideas, Suggestions and Inventions Committee

According to the Patents Act of 1978, "A patent may…be granted for any new invention which involves an inventive step and which is capable of being used, or applied in trade or industry or agriculture".[23] An invention had to be innovative, novel and relevant. This was little different to the criteria established for an invention by the Inventions Committee during the Second World War: "All inventions should contain an element of surprise even if the only surprise is that it was never thought of before…the recognition of the need is more than half of any invention; the meeting of the need is generally the easiest part of the matter".[24] It was not then a matter of simply developing a device to serve a purpose; the utility of such a device had to be ascertained and key was its innovative nature. This, then, would be the criteria applied by members of the Committee to the many inventions submitted by both civilians and military personnel over the course of the War.

[23] Republic of South Africa Government Gazette, No 57 of 1978: Patents Act, 1978, Cape Town, 17 May 1978, p20.

[24] W.E. John, DGTS, "Re War Inventions", 28 August 1944, File Q(T) 19/234, Vol 1: Inventions—Destruction of Mine Fields by Major A.S.J. Du Toit, KMG, Group 27, Box 143, Department of Defence Archives (hereafter DOD).

Initially, for the first two years of the War, there had not been sufficient co-ordination in the reception and assessment of inventions submitted to the UDF by both civilians and military personnel. While most submissions were to be addressed by the Director General of War Supplies (DGWS), they were also considered by various other departments. The DGWS would then deliberate on the inventions and submit them to the various sectors of the UDF to whom they were likely to apply. In addition, provision had to be made for the compensation due to inventors for feasible and successful ideas.[25] The DGWS however was reluctant to oversee inventions, believing their role related to the "production" of technology and war equipment and not to its introduction. As a result, the UDF liaised with the British Ministry of Supply (MoS), requesting from the latter that the South Africans be permitted to submitted inventions to them as there was no existing body in the Union that could "effectively deal" with the inventions submitted. The response was favourable with the MoS sending to the UDF two lists—the first indicating potential areas of "interest" and another with the proposals that they had already received and in which they were no longer "particularly interested". As a result, approximately thirty suggestions received by the UDF were subsequently sent to the MoS in London in 1941 for their consideration.[26]

By 1942, the Director General of Technical Services (DGTS) was willing to take on the role of overseeing and co-ordinating the submission of inventions—although it would still submit those ideas which it lacked the capacity to address to the MoS.[27]

Six months after the new structures were put in place, P.A. Leon, who had served as the head of the Committee in its earlier incarnation, reported on its progress since its inception the previous year. Central to Leon's report was a criticism of existing efforts—or the lack thereof—to draw attention to the work of the Committee by the UDF and so invite

[25] K.R. van der Spuy, Brigadier General DGTS, "Force Orders UDF All Sections", 27 October 1943, File Q(T) 19/234, Vol 1: Inventions—Destruction of Mine Fields by Major A.S.J. Du Toit, KMG, Group 27, Box 143, DOD.

[26] P.A. Leon, "Report by the Chairman of the Ideas, Suggestions and Inventions Committee of the Johannesburg Defence Force Liaison Committee for the Year Ended 30th June 1942", File Q(T) 19/234, Vol 1: Inventions—Destruction of Mine Fields by Major A.S.J. Du Toit, KMG, Group 27, Box 143, DOD.

[27] Leon, "Report by the Chairman".

the submission of ideas from both military and civilian circles. He emphasized that the Committee itself was not an outlier but was very much in line with efforts made in other Allied countries, including the United States as well as South Africa's fellow Dominion, Australia. The National Inventors' Council in the United States had apparently received 45,000 submissions, many of which were actually feasible and implemented.[28] Similarly, as Edgerton shows, just two months after the outbreak of the War, the MoS had received 20,000 submissions and approximately 50,000 by June 1941, the overwhelming number of which came from civilians. The average rate of submissions was 1,500 per week but this rose during times of exceptional crisis such as the Battle of Britain and the Blitz.[29]

As Leon pointed out, the Dominion, Australia, was also actively seeking out new ideas rather than displaying a "conservative distrust of anything new and unusual"—a possible criticism of the less than enthusiastic attitude of the UDF—and this was to be a defining feature of the Australian Army Inventions Board under the leadership of a South African, Richard van der Riet Woolley.[30] Woolley had actually been born in England in 1906, the son of Paymaster Rear Admiral Charles Woolley. His maternal grandfather however was Frederic J. van der Riet, a magistrate in Simonstown and his family had close links with scientific and political figures in South Africa, including Jan Smuts and pioneering geologist, A.W. Rogers. Woolley was a teenager when the family emigrated to South Africa and subsequently studied science at the University of Cape Town, with Basil Schonland one of his lecturers in Physics. Clearly gifted, he was awarded his MSc at the age of nineteen. During further study at Cambridge, he developed a strong interest in astronomy and, just after the outbreak of the Second World War, he was appointed Commonwealth Astronomer at the Commonwealth Solar Observatory near Canberra. The Observatory was soon to become part of the war effort as a munitions factory producing optical glass. Woolley then headed the Army Inventions Directorate, his role to co-ordinate the "[sifting through of] thousands of suggestions for winning a war".[31]

[28] Leon, "Report by the Chairman".

[29] Edgerton, *Warfare State*, p350.

[30] Leon, "Report by the Chairman".

[31] William McCrea, "Richard van der Riet Woolley, 24 April 1906–24 December 1986", *Biographical Memoirs of Fellows of the Royal Society*, 922–982, Vol 34, December 1988, pp924–926, 929, 932–934.

For Leon however, the efforts of the UDF were falling far short of that of the other Allied countries. In contrast to the "thousands" of submissions submitted in the United States, Britain and Australia, the initial publicity when the Committee was launched resulted in a flurry of 306 submissions in the first two months. The subsequent year tendered a total of just 363, the first three hundred as a response to a newspaper article drawing attention to the work of the Committee. From the relatively paltry number of submissions, Leon concluded that it was initially due to the poor system of organization in dealing with the proposals. Ideas were usually exploratory rather than complete and Leon believed that prospective inventors submitted them with the expectation that they would be permitted to work alongside the "technical members" of the Committee to "experiment further and to endeavour to perfect their inventions". On the part of the Inventions Committee, they had not been allocated the necessary resources by the UDF in order to fulfil the expectations of members of the public. While they assiduously conducted interviews with prospective inventors they were not equipped with the "facilities" such as the need to construct models for instance that would render their deliberation as anything other than "theoretical". The situation was worsened by the initial lack of co-ordination of reception of suggestions and their forwarding to the relevant departments that had bedeviled the early years following the establishment of the Committee. This was also a factor that prevented the Committee from making widespread appeals for ideas to the general populace. Perhaps most telling, however, was Leon's indictment of the UDF: "This Committee has had to contend with the lack of enthusiasm on the part of the military authorities to stimulate the submission of ideas".[32] Leon's report then emphasized what was apparently a lack of interest on the part of the UDF in the potential use of invention and innovation in contrast to that of the other Allied countries—and to the active role played by their citizenry.

At the time of his writing, the country was undergoing its own war crisis. Allied forces had experienced a massive setback in Tobruk in North Africa with thousands of South African troops taken prisoner. There was a decline in recruitment attributed to "war weariness" with the UDF eventually compelled to initiate changes in the tone of recruiting attempts and to make greater provision for greater inclusivity in the war effort—and in

[32] Leon, "Report by the Chairman".

the UDF. This was essential in a country confronted with an ambivalent response to participation in the War.[33] The work of the Ideas, Suggestions and Inventions Committee could conceivably then be placed within a context of increasing support for the war effort—at a time when this was waning—by encouraging the submission of inventions by citizens.

It was perhaps in response to the perceived lack of official attention given to the work of the Committee that in late December 1942, Brigadier General Trollip of the DGTS sent a memo reiterating the establishment of an "Inventions Committee" to address the "Ideas to Help with the War Effort". Accompanying the memo was a list of guidelines for the inventions sought by the UDF that he recommended be circulated regularly every three months, thus encouraging military personnel to submit their ideas.[34] For the purposes of ideas and inventions, it was not just the physical fitness of the UDF soldier that was key to the war effort but "the intelligence and brain power of the forces" and any member of the UDF regardless of rank was exhorted to lend their brain to the war effort in terms of making "suggestions...which may help to win the war". Special acknowledgement was made of the practical experience of serving men who had acquired the most familiarity with the equipment and weapons of war and were thus considered to be in the "best position to conceive ideas for improvement in method or design with a view to increasing efficiency".[35] A subsequent provision would stress that, while invention and innovation were to be stimulated, the focus of submissions should be on those of a "warlike nature" that would make an impact on the "Fighting Forces".[36] Simultaneously, the submission of ideas necessitated drawing attention to the shortcomings of existing equipment. While the DGTS encouraged ideas that "[built] on existing foundations"

[33] Refer to Suryakanthie Chetty, "Our Victory was Our Defeat: Race, Gender and Liberalism in the Union Defence Force, 1939–1945" (University of KwaZulu-Natal, Unpublished PhD Dissertation, 2006).

[34] B.G. Trollip, DGTS, "Suggestions, Ideas and Inventions", 29 December 1942, File Q(T) 19/234, Vol 1: Inventions—Destruction of Mine Fields by Major A.S.J. Du Toit, KMG, Group 27, Box 143, DOD.

[35] B.G. Trollip, "Draft Routine Order: Ideas to Help the Union's War Effort", File Q(T) 19/234, Vol 1: Inventions—Destruction of Mine Fields by Major A.S.J. Du Toit, KMG, Group 27, Box 143, DOD.

[36] Deputy Chief of Staff, "Experimental Services and Inventions", 16 January 1943, File Q(T) 19/234, Vol 1: Inventions—Destruction of Mine Fields by Major A.S.J. Du Toit, KMG, Group 27, Box 143, DOD.

rather than seeking to re-invent the wheel, it also made provision for "frank criticism" of existing technology, even if the criticism was "not agreed with".[37] The submission of inventions then, potentially offered some leeway to servicemen (and women—although this was not explicitly stated) that permitted the exercise of opinion and expertise outside the hierarchical structure.

A set of procedures was thus put in place for military personnel to address the submission of inventions. For members of the UDF struck with inspiration, they were expected to introduce concisely their ideas to their commanding officers who would in turn forward this to the DGTS. If military personnel were made aware of civilian inventions, these too had to be sent to the DGTS which was to be the organization in the UDF responsible for the assessment of the inventions.[38] As this was part of military bureaucracy, no procedural stone should be left unturned and proposals "should be in duplicate in double space typing and with a two-inch margin on the left side of each page".[39] The inventor should then timeously be informed as to whether the invention met the criteria i.e. "whether the idea is novel and is required". Should the invention do so, the DGTS then undertook the "development" and the "testing" of the invention. The entire process from the initial submission of the idea was to be kept confidential but if the invention proved to be of benefit to the UDF, the inventor was then eligible for monetary compensation. If a patent had to be issued, it would be done so in the name of the inventor while the UDF would retain the right to use the invention. Once the War had ended, the inventor could then benefit from the "commercial exploitation" of the invention.[40]

In any given month, the Inventions Department was responsible for:

> Scrutinising some 150 patent specifications
> Notifying departments of technical information contained therein and useful for the war
> Interviewing some 35 inventors

[37] Trollip, "Draft Routine Order".

[38] R.O. 1057—Inventions, 22 November 1943, File Q(T) 19/234, Vol 1: Inventions—Destruction of Mine Fields by Major A.S.J. Du Toit, KMG, Group 27, Box 143, DOD.

[39] Deputy Chief of Staff, "Experimental Services and Inventions".

[40] R.O. 1057—Inventions.

> Dealing with correspondence relating to some 38 suggestions[41] sent by post. The files in this connection already number well over 1,000. Constructing, finishing and arranging for testing of any necessary models.[42]

Further duties included determining awards to inventors where appropriate and Brigadier General Van Der Spuy recommended the establishment of a separate committee to consider the awards given for inventions "of a direct Military, Air Force or Naval character" submitted by either civilians or members of the UDF. Inventions would only be considered however if they were "genuine" and not simply a modification of an existing device already in use.[43] Important too, was the preservation of secrecy, especially for those ideas that would be developed and utilized. In this, the Undersecretary for Defence took into account the rights of the inventor. Should a patent be registered, it would be done so in the name of the inventor but confidentiality was to be maintained. A record should also be kept of when the invention was suggested and when it was first used so as to forestall any later claims by a "third party".[44]

Upon submission of an invention then, the inventor had to sign an agreement that spelled out both his rights in terms of the patent application as well as the rights of the state to the use of the invention. The inventor was allowed either before or even after they had submitted the invention to the UDF for consideration, to "file a provisional patent application" at the Union Patent Office. They were not however permitted to make use of an external agent in this regard until written permission had been received by the Ministry of Defence if it was convinced that this was not a violation of secrecy. Similarly, inventors were prevented from divulging their inventions to any other parties as well. A central

[41] An oddly specific number considering that Van der Spuy was detailing a monthly "average".

[42] Van der Spuy, "Force Orders UDF All Sections".

[43] Van der Spuy, "Force Orders UDF All Sections".

[44] Undersecretary, Defence Secretariat, "Regulations – Patents and Inventions", 17 September 1943, File Q(T) 19/234, Vol 1: Inventions—Destruction of Mine Fields by Major A.S.J. Du Toit, KMG, Group 27, Box 143, DOD. This would be the unfortunate situation in which Major A.S.J. Du Toit found himself in relation to the invention of the flail tank. Please refer to Chapter Four.

component of the agreement was the potential application of the invention "by the Union Government or her Allies and their contractors and others authorized by said Governments..."[45] The agreement then revealed not just the concerns regarding national security during a time of war but also South Africa's relationship with the Allies in terms of the sharing of technological innovation—that would be reflected in the actual inventions.

The relationship between South Africa and the Allies was also envisioned by the DGTS as one where there was the free exchange of technology with the objective of winning the War. The Secretary of Defence drew attention to an agreement made between the United States and Britain in relation to patent rights as it applied to technology used in war. This was so that, should the UDF wish to make use of "American patents for defence or for other war purposes", their legal position was clear.[46] In August 1942, less than three months before the United States officially entered the War after the Japanese Attack on Pearl Harbour, an agreement was formulated between the United States and Britain. The agreement stated that, based on "mutual aid in the prosecution of the war against aggression", the governments of the United States and Britain should allow access to each "patent rights, information, inventions, designs or processes" as required for the duration of the War.[47] Just as significant were the patents that were "enemy-owned or controlled" and again, the UDF took note of the American experience where these patents for inventions or "processes" had to be made available to the United States if they were "in the public interest" and necessary to the war effort. With this in mind, the office of the Alien Patent Custodian recommended "[taking] possession of all enemy-owned or controlled patents" and these patents would, if required, be made available to American war

[45] "Union Defence Force: Regulations Regarding Patents and Agreements", File Q(T) 19/234, Vol 1: Inventions—Destruction of Mine Fields by Major A.S.J. Du Toit, KMG, Group 27, Box 143, DOD.

[46] Secretary for Defence, "Patent Interchange Agreement between the United States and the United Kingdom Government", 28 November 1942, File Q(T) 19/234, Vol 1: Inventions—Destruction of Mine Fields by Major A.S.J. Du Toit, KMG, Group 27, Box 143, DOD.

[47] "Patent etc. Interchange Agreement", 24 August 1942, File Q(T) 19/234, Vol 1: Inventions—Destruction of Mine Fields by Major A.S.J. Du Toit, KMG, Group 27, Box 143, DOD.

industry.[48] Inventions—and the patents associated therewith—were thus a means of expressing nationalism, Allied alliances—and enmities. Simultaneously, despite the UDF's concern with patents (and protecting both the UDF and the inventor), Edgerton suggests that patents are not of great import—there are differences across states in terms of patent laws, they give little inkling of the actual value of an invention; and they are employed haphazardly by inventors, if they are employed at all. Patent law is ultimately a piece of legislation not utilized to its fullest extent—yet during war and in terms of agreements between Allies, they were nevertheless an integral part of the promotion and sharing of invention.[49]

With the streamlining of the Inventions Committee, the DGTS subsequently enlisted the expertise of a consulting engineer, W.E. John to head the Inventions Committee in the DGTS.[50] The choice of John as the chair was based on two key criteria required by the DGTS in relation to the submissions of inventions; the first was a technical ability to assess the invention and the second was a knowledge of and familiarity with patent law so as to safeguard the interests of the UDF and the inventor.[51] John had served in an informal capacity as an advisor to the previous Defence Liaison Committee and was already on the Reserve of Officers. He had worked as a technical advisor with regard to inventions in both the United States and Britain as well as an Examiner of Patent Agents in the Union.[52] He was also a member of various technical professional bodies: the Institution of Mechanical Engineers and the Royal Aeronautical Society in Britain as well as the Institute of Engineers in South Africa.[53]

[48] Leo T. Crowley, Alien Property Custodian, "Statement Concerning Enemy-Held Patents and Trade-Marks", File Q(T) 19/234, Vol 1: Inventions—Destruction of Mine Fields by Major A.S.J. Du Toit, KMG, Group 27, Box 143, DOD.

[49] David Edgerton, *The Shock of the Old: Technology and Global History Since 1900* (London, Profile Books, 2008), p200.

[50] Van der Spuy, "Force Orders UDF All Sections".

[51] DTD, "Ideas and Inventions", File DTS 405, Vol 2: W.E John Esq, Consulting Engineer and Patent Agent, KMG, Group 27, Box 166, DOD.

[52] Letter by W.E. John to Brigadier General K. van der Spuy, 17 April 1942, File DTS 405, Vol 2: W.E John Esq, Consulting Engineer and Patent Agent, KMG, Group 27, Box 166, DOD.

[53] Letter by W.E. John to Colonel Miller, Technical Services, 24 April 1942, File DTS 405, Vol 2: W.E John Esq, Consulting Engineer and Patent Agent, KMG, Group 27, Box 166, DOD.

As well as his impressive credentials, John also had a clear policy for dealing with the submission of inventions and an optimistic view of the importance of invention. Acknowledging that, of the many submissions, "only one idea in a thousand" could be expected to have any merit, he nonetheless pointed out that that one single idea had the potential to justify all the effort made on the part of the Inventions Committee. To be able to find this potentially groundbreaking idea then required a person with longstanding technical experience and with sufficient technical secondary material on which to draw, as well as legal texts related to patent law. John spelled out his own procedure for dealing with potential inventors: a meeting where the inventor described his invention and presented any illustrative material such as a "model and sketches;" John's subsequent research into the feasibility of the idea; and a rudimentary "specification" which is then discussed with the inventor. The final stage was the submission of the registered patent to the Patent Office. With regards to an invention submitted to the UDF, John envisaged the potential conflict between inventor and Committee in terms of the development of the idea. His experience had suggested that the inventor prioritized enthusiastically the "[alteration and improvement]" of the invention before the initial testing had been carried out. Once the invention was submitted, the input of the inventor was however no longer required unless "exceptional difficulties" presented.[54] Perceiving the current conflict as an "engineer's war",[55] John's recommendations clearly resonated with the DGTS. With its new chair and a commitment to having its work publicized, the Ideas, Suggestions and Inventions Committee was poised to receive inventions from all over the Union.

From the Crackpot to the Credulous

The formation of various invention committees during the Second World War can be contextualized by a popularization and perhaps democratization of science and technology that acknowledged the potential of the amateur. In 1855 Samuel Beeton launched the inaugural *The Boys Own Magazine* that would run for the next two decades. Aimed at

[54] Letter by W.E. John to Colonel Miller.

[55] Letter by W.E. John to Brigadier General K. van der Spuy, 17 June 1942, File DTS 405, Vol 2: W.E John Esq, Consulting Engineer and Patent Agent, KMG, Group 27, Box 166, DOD.

pre-adolescent and teenage boys, Beeton described the periodical as "Tales of Adventure, Stories of Heroism and Courage" that featured excerpts on history and natural history, sport and outdoor activities and inspiring accounts of the boyhoods of "great men". The magazine wrapper depicted instruments such as the recently invented camera, a telescope along with various other "mathematical and scientific instruments", and the young readership was introduced to figures that included the inventors James Watt and Benjamin Franklin, along with the romanticized explorers such as James Cook. In similar vein, boys were exhorted to carry out basic "mechanical experiments" and introduced to "The Tools of War". The latter, though, were decidedly less modern and included a long-sword and bow and arrow. The enormous popularity of the magazine—with a readership of some forty thousand across the British Empire—led to a second series published from 1862 that, at one hundred pages, was three times the length of the original magazine. In the new format, brief excerpts on "useful knowledge" were juxtaposed with lengthier expositions of science and history. Other editors would soon emulate Beeton's publication.[56]

Fifty years later, in January 1902, Henry Haven Windsor published the first issue of *Popular Mechanics*. With early editions running up to six hundred pages, the periodical was touted as being "written so you can understand it" and was a font of information on the latest technologies and inventions, including the use of an internal combustion engine on a locomotive rather than a steam engine, experiments with flight, airships and aeroplanes. An article in 1928 described the possible "inventions of the future" that included the use of solar and wave power as well as "a more flying-friendly future" with a helicopter in every "backyard".[57] In 1908, came the publication of *Amateur Mechanics*, a compilation of articles from *Popular Mechanics* that was aimed at "young and old who like to make things". As with the descriptor, "popular", "amateur" too, emphasized the accessibility of mechanics and mechanical invention. Along with useful hints in keeping animals away from trash cans and rust removal,

[56] Louis James, "'Now inhale the gas': Interactive Readership in Two Victorian Boys' Periodicals, 1855–1870", *Victorian Periodicals Review,* 64–80, Vol 42, No 1, 2009, pp64–69.

[57] "Our Favourite Stories from 120 Years of 'Popular Mechanics'", 11 January 2022, https://www.popularmechanics.com/about/a38726786/our-favorite-stories-from-120-years-of-popular-mechanics/, accessed 2 March 2024.

there were instructions for experiments in electricity, in making a wireless telegraph, an arc light and broadcasting music from a phonograph. For younger readers, were instructions on the creation of illusion—"magic" or "parlour" tricks—and the construction of toboggans. An article detailing the building of a telescope was titled "Farmer's Boy Builds a Telescope and Discovers a Comet" and was an articulation of the accessibility of science and technology—although the readership was presumed to be male.[58]

While these periodicals were largely factually-based, the science fiction writings of figures such as H.G. Wells and Jules Verne also inspired the popular imagination with the possibilities—and pitfalls—of science and technology.[59] Therefore, echoing the BIR a generation earlier, W.E. John of the Ideas, Suggestions and Inventions Committee, recommended the careful consideration of every idea put forth by a member of the public: "...every suggestion put forward, no matter how crude or fantastic, must be investigated and answered briefly perhaps, but with a full and obvious appreciation of the proposition".[60]

Over the course of the first year of the War some 173 ideas were submitted to the DGWS Committee and these mainly fell under "Anti-Tank Devices", "Aircraft Devices", "Bombing Improvement Suggestions" and "Firearm and Projectile Devices" with the majority classified as "General". One prospective inventor had even submitted an idea for "perpetual motion". Yet all ideas were accorded due diligence with only one showing promise, "a cathode-ray ocillograph [sic]" submitted by geophysicist Oscar Weiss that could potentially be used to detect aircraft up to sixty kilometres away, "through the detection of sub-audible vibrations". Members of the panel believed that the device could possibly be used to determine either the type of aircraft or the number of approaching aircraft and the idea was forwarded to the Director of War Supplies, H.J. van der Bijl for his consideration, and experiments were subsequently

[58] Refer to *Amateur Mechanics* (Chicago, Popular Mechanics, 1907).

[59] Paul Kennedy, *Engineers of Victory: The Problem Solvers Who Turned the Tide in the Second World War* (Penguin, 2014), pp55–56.

[60] Letter by W.E. John to Colonel Miller.

carried out.[61] The date of Weiss's submission is significant as it was during this period that the Battle of Britain was taking place, making the detection of enemy aircraft vital and it was the cavity magnetron radar that would come to the fore, giving the Allies a significant advantage. During the Battle of Britain, radar was used to detect incoming German aircraft, allowing sufficient time to scramble British fighters to counter German bombers.[62] Weiss had likely been drawing on the work of Robert Watson-Watt who, in 1933, had published work on the use of the "Cathode-Ray Oscillograph". Just two years later, he would become the innovative figure in the development of radar with the use of radio waves to detect objects and was essential in setting up the radar stations in Britain that would eventually thwart the Germans.[63] Weiss, unfortunately, was treading on already established territory yet his submission—and the attention paid to it—demonstrated the convergence of utility and invention.

A suggestion for jet propulsion also demonstrated similarities across ideas—and the similar treatment of them. In 1944, Ronald Bullock of Florida in the Transvaal proposed improvements to a jet propulsion system for aircraft that would result in the nozzle of the jet being positioned in ways to improve the "aerodynamic performance of the wing". The wing of the aircraft would itself be adapted to allow for a "cooling chamber" to deal with the gases emitting from the jet. The aim of this detailed proposal was to create a faster aircraft by increasing "lift and thrust". Over the course of the year, Bullock's proposal came in three parts, each an adaptation or improvement on the one before. Accompanying the proposal was a detailed consideration of the physics and mechanics of flight and the envisaged performance of his invention, along with detailed diagrams[64] (Fig. 2.1).

[61] J.H. Wessels (DGWS), "First Report on Suggestions and Inventions", 21 August 1940, and letter by J.H. Wessels to O. Weiss, "Invention", 23 July 1940, File A3: Inventions Outstanding—Under Investigation as at 20-8-1940, KMG, Group 27, Box 178, DOD.

[62] Kennedy, *Engineers of Victory*, pp92–93.

[63] "Sir Robert Watson-Watt, Engineer, Inventor, And Pioneer Of RADAR", https://engineeringhalloffame.org/profile/robert-alexander-watson-watt, accessed 2 March 2023.

[64] Ronald William Bullock, "Improvements in Jet-Propelled Aeroplanes", "Jet Sustentation of Aircraft (Jet Propelled)" and "The Nozzle and Aerofoil Jets", File 462: Jet Propulsion—Inventor: R.W. Bullock, KMG, Group 27, Box 169, DOD.

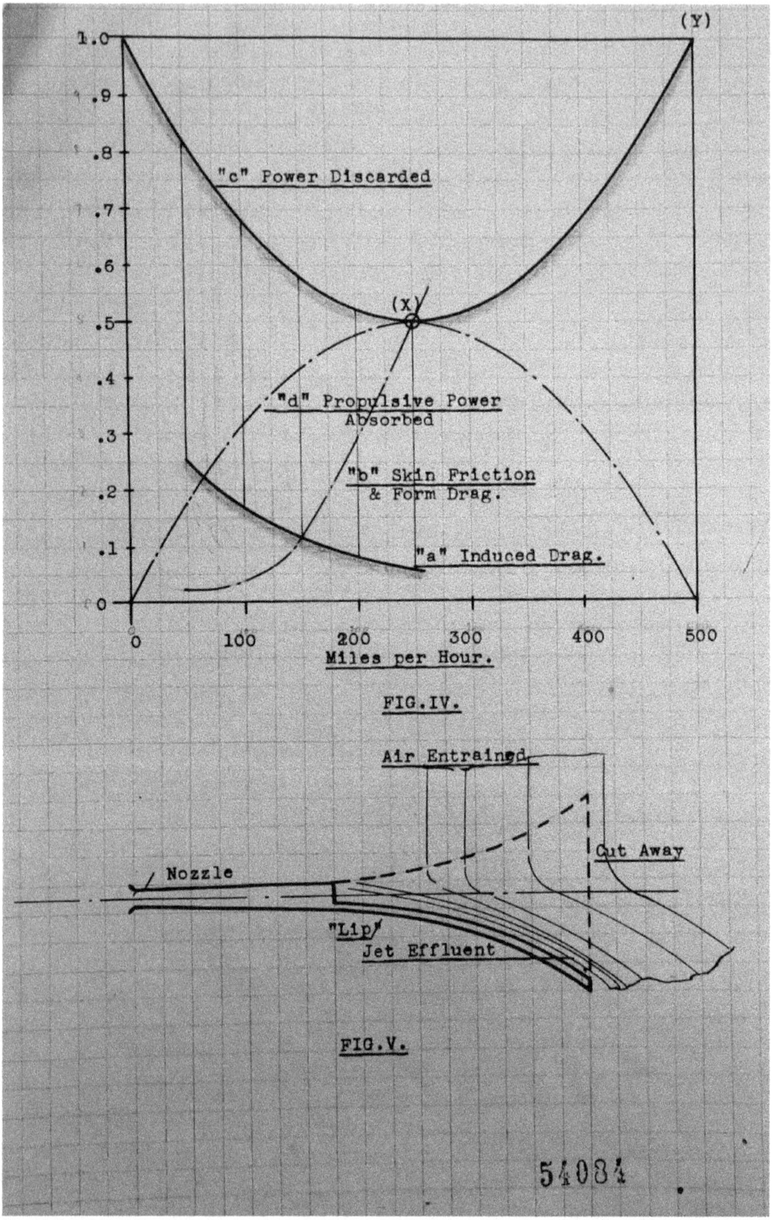

Fig. 2.1 Ronald William Bullock, "Improvements in Jet-Propelled Aeroplanes", "Jet Sustentation of Aircraft (Jet Propelled)" and "The Nozzle and Aerofoil Jets", File 462: Jet Propulsion—Inventor: R.W. Bullock, KMG, Group 27, Box 169

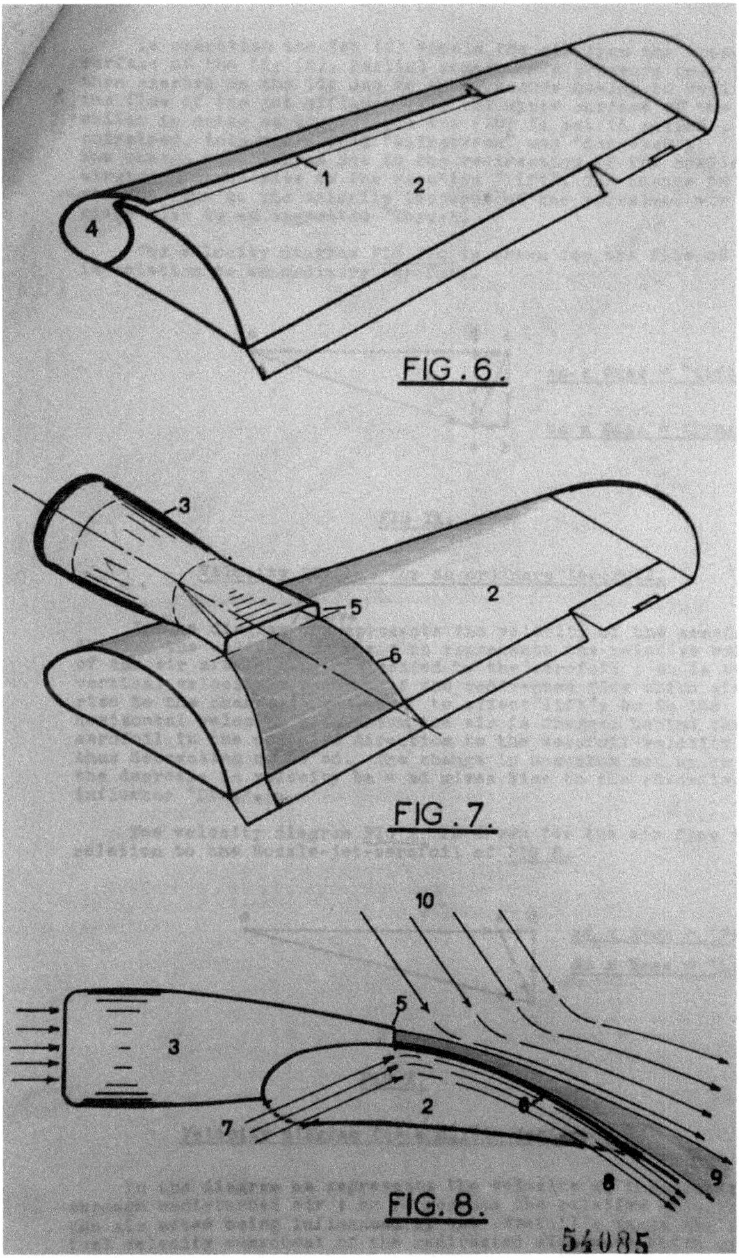

Fig. 2.1 (continued)

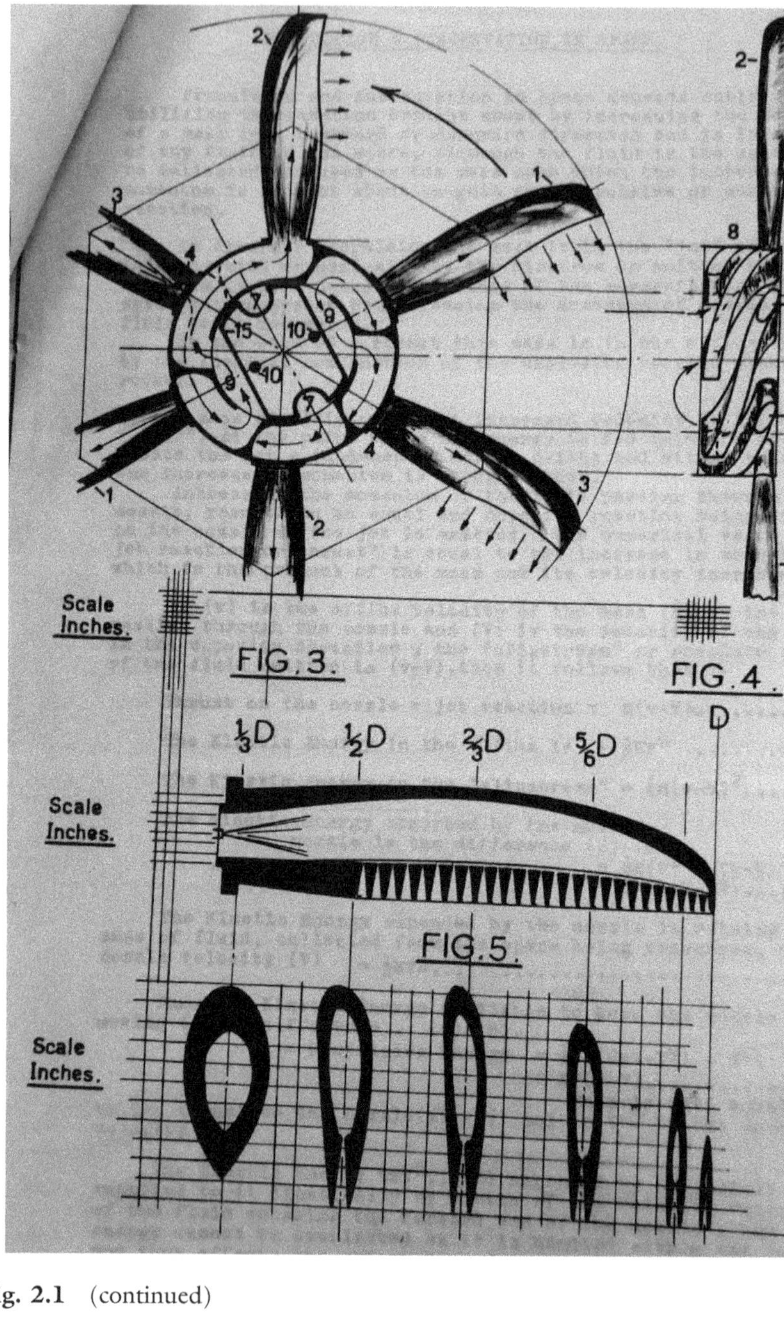

Fig. 2.1 (continued)

Bullock's proposal, running to several typed pages in total, looked formidable and he was willing to share it with the Allied countries during the War, provided that his patent rights were protected. The DGTS subsequently submitted it to the MoS in Britain.[65] Bullock's design was given due consideration at the Ministry of Aircraft Production's Department of Scientific Research, however the conclusion was that the submission was "not new" and that some of the effects on drag and lift were not as Bullock envisaged them with the "disadvantages" of his design "[outweighing] the advantage".[66] In Bullock's file too, there is a clipping of a newspaper article mentioning Frank Whittle, the man who invented the jet engine.[67] Bullock's jet propulsion system was thus part of an already existing conversation regarding Britain's adoption of the jet engine. Of humble origins, Whittle had served as an apprentice in the RAF before being promoted to test pilot where he developed numerous proposals for new inventions and as a pilot, tested devices such as the catapult on aircraft carriers and a device to keep aeroplanes afloat if they crash landed in the ocean. Whittle also began formulating an idea for a jet propulsion engine which had significant military potential—advance warning radar systems combined with the superior speed of a fighter equipped with a jet engine could give the British a very effective defensive system when under aerial attack. Between 1936 and 1937, Whittle and his engineering team built "the world's first working gas turbine engine". However, the British were slow to capitalize on this advantage and Whittle's work was not given the support required. On 26 July 1944, an RAF pilot reported being attacked by a Messerschmitt Me 262 that flew at least 160 km faster than his own Mosquito. This was the prototype of a jet-propelled fighter, developed and then used by the Germans.[68] Bullock's proposal then was not new but also occurred within a context

[65] Technical Services Section, "Re Jet Propelled Aeroplanes by R.W. Bullock", 21 March 1945, File 462: Jet Propulsion—Inventor: R.W. Bullock, KMG, Group 27, Box 169, DOD.

[66] N.F. Baker, "Jet Propelled Aeroplanes", 7 May 1945, File 462: Jet Propulsion—Inventor: R.W. Bullock, KMG, Group 27, Box 169, DOD.

[67] "16 Engineers to Leave Jet Plane Firm", *Mail*, 16 April 1946, File 462: Jet Propulsion—Inventor: R.W. Bullock, KMG, Group 27, Box 169, DOD.

[68] Duncan Campbell-Smith, *Jet Man: The Making and Breaking of Frank Whittle, Genius of the Jet Revolution* (London, Head of Zeus Ltd, 2021), pp65–66, 97, 177, 402.

where the potential of the original proposal had been realized too late. It demonstrated that the inherent utility of an invention was not necessarily self-evident.

It was, however, a utilitarian invention, one that could by no means be termed ambitious as was the case with jet propulsion, and one that demonstrated the experience of the soldier in identifying a need and developing a means to address it that would prove to be successful. Corporal Killoran's idea for a deflector plate for a Bren Gun was a brief twelve-line paragraph. In it, he described a plate that would be attached to the right side of the gun that would deflect the spent cartridges away from the gun. In the current state, the shells fell to the ground, raising dust which could potentially jam the mechanism of the gun or obscure the vision of the gunner—all potentially hazardous in combat. It was a simple suggestion and Killoran believed that it would be extremely cheap to produce.[69] Killoran's solution struck the right note, a prototype was developed by the Armoury and tests carried out where it was found that the plate had to be positioned at a certain angle to ensure a deflection of 45 degrees that would have the further advantage of deflecting the hot cartridges away from the gunner's body.[70] An initial estimate of a thousand deflectors was subsequently recommended for production at a cost of 1/- each.[71] The success of the invention was such that the UDF forwarded the design to its Allies: New Zealand, Australia, India, Canada, the United States and Britain.[72] Successful inventions, however, that were actually put into production were few and far between.[73]

While Killoran's deflector was a way of compensating for the difficulties of warfare in North Africa, the harnessing of sand was also a potential weapon. Robert Booth was already employed in armoured car

[69] J.A. Killoran, File DTS/INV 338: Bren Gun—Proposed Improvements to—Deflector for Empty Cases: Cpl J.A. Killoran, KMG, Group 27, Box 163, DOD.

[70] ADTD, "Report", File DTS/INV 338: Bren Gun—Proposed Improvements to—Deflector for Empty Cases: Cpl J.A. Killoran, KMG, Group 27, Box 163, DOD.

[71] DCS, "Bren Gun Cartridge Case Deflector", 6 June 1942, File DTS/INV 338: Bren Gun—Proposed Improvements to—Deflector for Empty Cases: Cpl J.A. Killoran, KMG, Group 27, Box 163, DOD.

[72] Refer to File DTS/INV 338: Bren Gun—Proposed Improvements to—Deflector for Empty Cases: Cpl J.A. Killoran, KMG, Group 27, Box 163, DOD.

[73] For more in-depth case studies of successful South African inventions, please refer to Chapters Four and Six.

manufacture in Johannesburg. Aware of the desert conditions—and the disruption and discomfort created by sandstorms—in North Africa, he suggested the creation of artificial sandstorms to wreak havoc among the enemy as the Allied armoured car divisions advanced. Booth recommended the installation of an air-cooled aircraft engine mounted on the back of the armoured car that would channel air around the car and so create an artificial sandstorm.[74] While Booth was subsequently assured of the interest that his submission had aroused, there was some dissent regarding its potential implementation. The submission was discussed at a meeting of the Committee held in September 1940 with the resulting view that, although it appeared to be "far-fetched and perhaps impracticable", compounded by the additional power required to drive the aircraft engine, it was nevertheless feasible to create a "moving wall of sand" in advance of the vehicle.[75] This tentative and qualified recommendation for "further investigation" was given short shrift by G. Ractliffe of the DGWS, "I beg to differ. The enormous horse power required...rules the whole scheme out".[76]

While the artificial sandstorm teetered on the boundary between a potentially feasible invention and a "farfetched" one, other suggestions were more easily dismissed by a rather more tolerant W.E. John. A slew of correspondence is evident between the very persistent W.A. Gaunt of the Natal Government Training College in Pietermaritzburg and the DGTS related to Gaunt's proposal of a detachable curved gun barrel[77] (Fig. 2.2).

Gaunt proved persistent in his development of the idea, claiming to have carried out his first "experiment" at the age of 12 with an air gun and "some curved piping". The pipe was fitted over the muzzle of the weapon and he was able to hit a target about four metres away. The curvature however reduced both the speed and hence the power of the

[74] Letter by Robert Booth, 3 September 1940, File B4: Mr Robert E. Booth—Artificial Sandstorms, KMG, Group 27, Box 178, DOD.

[75] Ideas, Suggestions and Inventions Committee Memorandum to Liaison Officer, 12 September 1940, File B4: Mr Robert E. Booth—Artificial Sandstorms, KMG, Group 27, Box 178, DOD.

[76] G. Ractliffe, "Letter from Mr Robert E. Booth—Artificial Sandstorms", 15 November 1940, File B4: Mr Robert E. Booth—Artificial Sandstorms, KMG, Group 27, Box 178, DOD.

[77] Refer to File 334/9: Detachable Curvature Barrel Extension for Guns—Inventor: W.A. Gaunt, KMG, Group 27, Box 163, DOD.

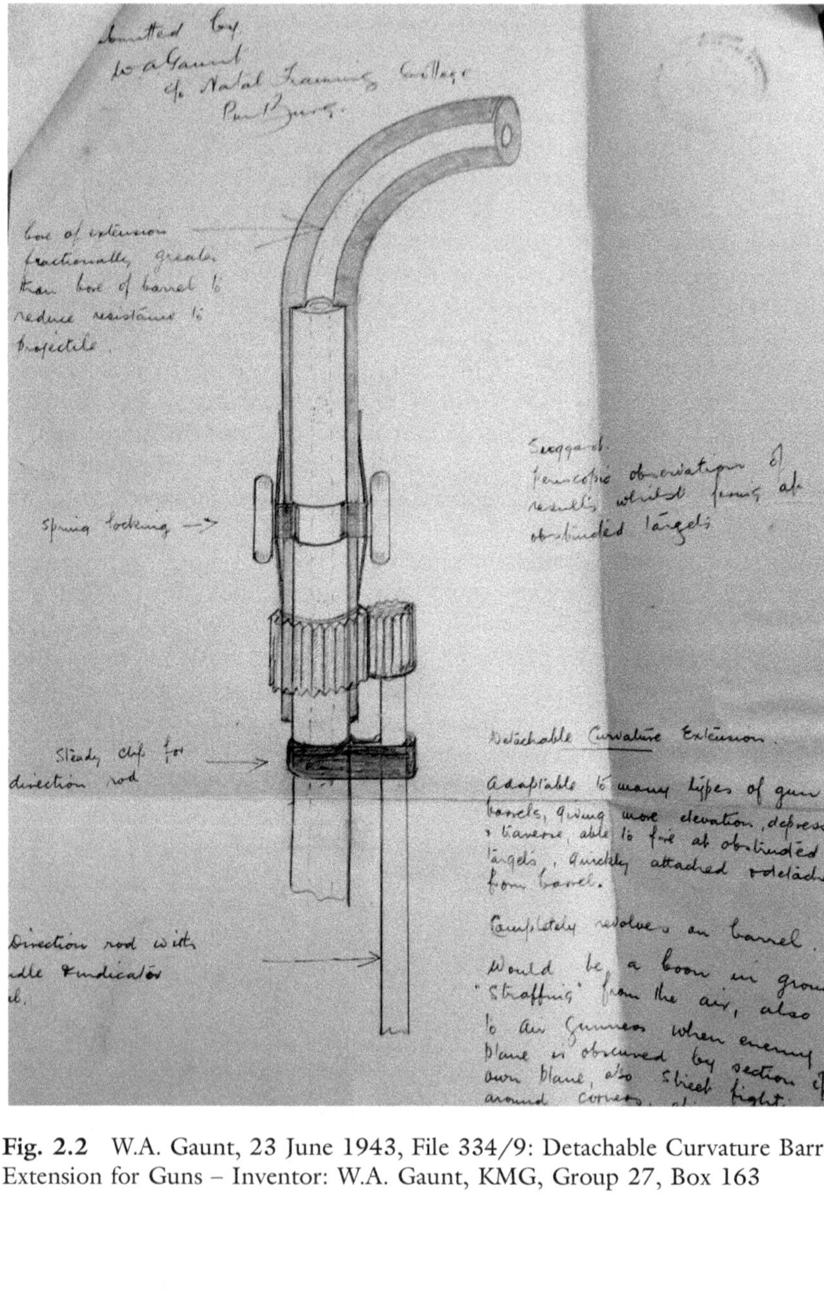

Fig. 2.2 W.A. Gaunt, 23 June 1943, File 334/9: Detachable Curvature Barrel Extension for Guns – Inventor: W.A. Gaunt, KMG, Group 27, Box 163

pellet. During the First World War, while stationed in East Africa he came across a curved pipe approximately 25 centimetres in length, fitted it onto his rifle and was able to hit a tree trunk less than 20 metres distant. The trunk was a little more than half a metre in diameter. By 1932, Gaunt's experiments had become less haphazard with the building of an "object resembling a gun" complete with curved gun barrel. At the time of writing in 1943, he had constructed the device depicted in his accompanying illustration. His belief was that there were no limits to the potential uses of his weapon in its utilization against the Axis and that he could certainly envision it being used in urban warfare to be fired "around street corners".[78] A month later, Gaunt received the standard rejection letter from John, thanking him for his submission but stating that the Committee did not consider it suitable for further development.[79] A standardized rejection proved insufficient deterrence and Gaunt replied with a "keen sense of disappointment", citing the expertise he had enlisted from a member of the Technical College—at which he was based—as well as a "professor" with "mathematical qualifications" who had considered his proposal feasible. As such, he wished the Committee to provide him with the scientific basis for their rejection of the proposal. He further inquired as to whether he could then submit his design to the United States or to Britain.[80] John replied that the Committee's "criticisms" could not be "furnished" and that the DGTS would take "no further action in this connection".[81] An unequivocal rebuff, it deterred Gaunt for all of two years before he wrote to the Committee again enclosing an article from the *Illustrated London News* describing a similar device invented by the Germans and stating that he could have developed an improved weapon under the "scientific guidance" of the DGTS.[82] John's response thanking

[78] W.A. Gaunt, 23 June 1943, File 334/9: Detachable Curvature Barrel Extension for Guns—Inventor: W.A. Gaunt, KMG, Group 27, Box 163, DOD.

[79] W.E. John, "Re Invention – Riffle [sic] Barrel", 5 July 1943, File 334/9: Detachable Curvature Barrel Extension for Guns – Inventor: W.A. Gaunt, KMG, Group 27, Box 163, DOD.

[80] W.A. Gaunt, 8 July 1943, File 334/9: Detachable Curvature Barrel Extension for Guns—Inventor: W.A. Gaunt, KMG, Group 27, Box 163, DOD.

[81] W.E. John, "Re Curved Gun Barrels", 19 July 1943, File 334/9: Detachable Curvature Barrel Extension for Guns—Inventor: W.A. Gaunt, KMG, Group 27, Box 163, DOD.

[82] W.A. Gaunt, 8 September 1945, File 334/9: Detachable Curvature Barrel Extension for Guns—Inventor: W.A. Gaunt, KMG, Group 27, Box 163, DOD.

Gaunt for his "continued interest" was tactful yet contained more than a hint of exasperation. It also held true to the spirit of inventions not presenting something completely new and ambitious, should there be existing and more efficient methods available:

> ...it is very much questioned whether the elaborate equipment necessary to spill bullets only 100 yds from the muzzle of the gun would be worth while. Probably an ordinary periscope and a few hand grenades would be considered more suitable for most emergencies. They would certainly be more effective and have the added advantage of being more readily available.[83]

Six months later saw the return of Gaunt with the submission of a newspaper article again describing the German gun which could potentially be adopted by the British. In such an event, Gaunt wished "rather reluctantly" to inquire that if the British made use of the gun, if the UDF would follow suit and, ultimately if he would be entitled to compensation given his earlier submission of the idea.[84] John replied in the affirmative to the first query and stated that Gaunt would be able to "make [his] claim" should such a situation arise. His concluding statement, however, was the last word on the subject: "It must be remembered however that the basic idea of a curved barrel for a gun to enable it to shoot around corners has been the subject of caricature for a very long time".[85] The letter was coincidentally dated the 1st of April.

While the curved gun barrel is reminiscent of the outrageous and far-fetched inventions associated with the comic books that had gained increasing popularity during the interwar years, the War nevertheless did take place within a context of competing ideologies—that brought with it a sense of conspiracy and espionage. A proposal for a planned "smooth bore small gun" suggested that the weapon was already utilized by the USSR as part of their secret arsenal. Typed in black and red ink with short, truncated sentences and a creative use of punctuation, the letter

[83] W.E. John, 12 September 1945, File 334/9: Detachable Curvature Barrel Extension for Guns—Inventor: W.A. Gaunt, KMG, Group 27, Box 163, DOD.

[84] W.A. Gaunt, 18 March 1946, File 334/9: Detachable Curvature Barrel Extension for Guns—Inventor: W.A. Gaunt, KMG, Group 27, Box 163, DOD.

[85] W.E. John, "Re: Curved Gun Barrels", 1 April 1946, File 334/9: Detachable Curvature Barrel Extension for Guns—Inventor: W.A. Gaunt, KMG, Group 27, Box 163, DOD.

stated that the inventor was in the process of developing the same, as evident in the accompanying sketch, however it was a Durban "byelaw" that "No firearm must not be discharged" [sic]. The missive concluded with the point that the country should emulate the Russians in the invention and use of the weapon, "If the others can do it. So should we Unless of course in this Capitalist Controlled Democracy. It will interfere with some one's Vested Interests?" The signatory was "TEIVOS"[86] (Fig. 2.3).

UDF officials very quickly deciphered "Teivos" as being "Soviet" in reverse. The letter, sent to the DGWS in Durban, was subsequently forwarded to the DGTS in Pretoria who gave the suggestion the same consideration as the others received, concluding that the weapon proposed was not practical, given the various trials that had already been carried out. Other than this brief and professional response to the invention itself, no comment was made regarding the eccentric nature of the submission.[87]

South Africa was not alone in entertaining inventions—and their enthusiastically persistent inventors—that possibly flouted the laws of physics. Described as an "indomitable race", inventors similarly

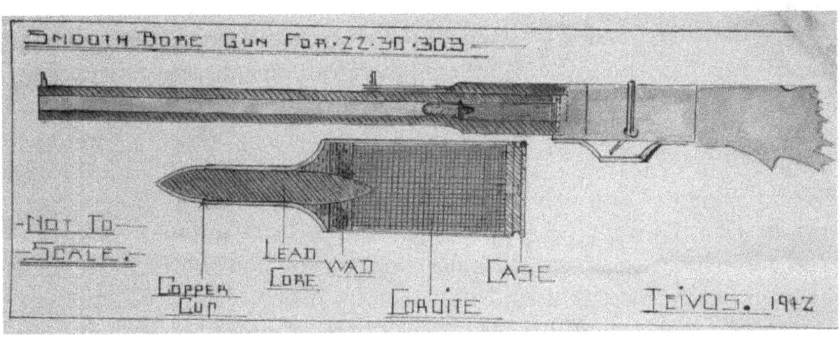

Fig. 2.3 P.C. Miller, DGTS, "Smooth Bore Gun – Invention", 12 May 1942, File 334/6: Smooth Bore Gun – Inventor: "Teivos" (Soviet), KMG, Group 27, Box 163, DOD

[86] "Teivos", Durban, 1942, File 334/6: Smooth Bore Gun—Inventor: "Teivos" (Soviet), KMG, Group 27, Box 163, DOD.

[87] P.C. Miller, DGTS, "Smooth Bore Gun—Invention", 12 May 1942, File 334/6: Smooth Bore Gun—Inventor: "Teivos" (Soviet), KMG, Group 27, Box 163, DOD.

bombarded the British with ideas that were classified according to their scientific merit, with some considered "cranks". The former could potentially be useful while the latter often presented nothing more than entertainment value: "flying tanks, jumping tanks, bayonets fixed to soldiers' boots…a gaseous and harmless compound which would solidify in the air as a gelatinous substance" containing the enemy, or as irreverently described, "Germans in aspic". It was a case of the mundane physical world being unable to keep pace with the inventive human mind.[88] While lighthearted—or exasperating for those duty-bound to contend with them—the image of the eccentric inventor was nevertheless a compelling one and a detraction from the abstract industrialization of war. It was also part of the narrative of the British at war.

The evacuation of Dunkirk was the outcome of the German blitzkrieg that swept through Western Europe. The British Expeditionary Force was rescued by all manner of craft from naval to the numerous fishing boats crewed by ordinary civilians. The story of heroism and of the indomitable will of regular Britons amidst military defeat became part of the British narrative and remembrance of the War, which would also include the Bitz and the Battle of Britain. As Edgerton argues, it presented an image of a country engaged in a defensive war and standing as the sole bulwark against Nazi aggression.[89] Related to this national mythology is the portrayal of the eccentric boffin. It is therefore unsurprising that it was in the immediate aftermath of Dunkirk that the Inspectorate of Anti-aircraft Weapons and Devices was established. Its popular moniker is reminiscent of the exploits of British public schoolboys: the Wheezers and Dodgers. The department was under Canadian Charles Frederick Goodeve who had been based at University College in London in 1939 and who had worked as an engineer. A relatively small department, the men who comprised the Wheezers and Dodgers would not have been out of place as staff in Q-Branch of Ian Fleming's James Bond novels. One of the recruits was a man who had invented "an unspillable ink-bottle" and an early attempt made by the department to protect naval shipping was the use of balloons to discourage enemy aircraft. The invention most associated with the

[88] "Many Inventions", *Rand Daily Mail*, 12 October 1940.

[89] David Edgerton, *Britain's War Machine: Weapons, Resources and Experts in the Second World War* (Penguin Books, 2012), ppl, 47–48.

Wheezers and Dodgers is the "Hedgehog"—the name, like the department in which it originated, emphasizing eccentricity and understating its lethal nature. The Hedgehog was an adaptation of and improvement on the depth charge used as an anti-submarine weapon. It comprised a number of mortars fired simultaneously ahead of the ship which increased the chances of hitting the target. Further, these were designed to explode on impact and not just when they had reached a certain depth.[90]

There are no definitive boundaries for invention. In a time when ideas were encouraged and the possibilities of science and technology seemed limitless, it was not always clear what constituted a feasible invention. Over the course of the War, the Allies feared that the Germans were in the process of developing "scifi"-type weapons such as an endothermic bomb—that would absorb energy rather than release it, as evident in conventional explosives—as well as a "death ray". And the Japanese actually did develop the latter using microwaves.[91] An invention that clearly lacked credibility could also be entertained if there was support for it. Geoffrey Pyke was Britain's Director of Programmes at Combined Operations during the Second World War and proved a font of inventions and schemes, including an oil pipeline across the English Channel. With a particular penchant for ice, Pyke envisaged ways in which it could be utilized as a tool of war. His moment of notoriety came with his proposal for an aircraft carrier made of ice. Observing that icebergs seemed impervious to projectile weapons and are less dense than water so were unsinkable, Pyke developed an idea for an aircraft carrier made of ice—that would be far more cost-effective than the conventional steel carriers. To compensate for the brittle qualities of ice, he suggested the addition of wood pulp to strengthen it. This would be named Pykrete in honour of its inventor. The scheme was supported by Pyke's chief, Louis Mountbatten and by Winston Churchill and, despite significant opposition, attempts were made to develop it in the remote—and icy—Canadian Rockies. Resources were poured into the project, which was ultimately abandoned due to, among other factors, its "comic book" overtones and

[90] Gerald Pawle, *The Secret War, 1939–1945* (London, George G. Harrap and Co. Ltd, 1956), pp21, 35–36, 123–124.

[91] Tom Shachtman, *Laboratory Warriors: How Allied Science and Technology Tipped the Balance in World War II* (New York, Perennial, 2003), pp259–260, 304–305.

the enormous amounts of steel required for refrigeration.[92] Invention was not based on the entirely novel but on adapting more effectively what already existed. Yet, on the other end of the spectrum, significant developments such as the jet propulsion engine could be marginalized.

The overwhelming number of suggestions—whether feasible or not—unsurprisingly related specifically to the War and weaponry and technology associated with it, however there were ideas that challenged the very notion of invention—and did not relate to technology at all. The Committee's title, "The Ideas, Suggestions and Inventions Committee", potentially allowed greater scope for the public to exceed conventional notions of invention, innovation and the novel.

In 1940, J. Cronje suggested the use of indigenous plants to serve as a source of food and water for South African troops in North Africa to strengthen them for their advance against the enemy. The plants suggested were *aalwyn* (aloe) and the "magawa" tree which was understood to mean *Kirkia wilmsii*. According to Cronje, the flowers of the aloe have a sweet-tasting water at the centre, which could be obtained for drinking by removing the stamen. The "magawa" tree, on the other hand, had a large underground tuber which was both edible and contained water. He therefore recommended that those servicemen in the north—especially those who were stranded such as pilots whose planes had been downed—make use of these natural sources of sustenance and also be aware of other useful flora.[93] Cronje's suggestion was given sufficient credence to be forwarded to the Department of Agriculture and Forestry for expert verification. This was not forthcoming—while not poisonous, the aloe plant described only flowered for short spells and the "nectar" was likely to "accentuate" rather than relieve thirst. As for the "magawa" tree, while its properties were as Cronje described, the tree itself did not occur in North Africa. Further, there was insufficient expertise regarding the flora of the region—their edibility could only be determined from "residents or natives".[94] Cronje's recommendation of the aloe or *aalwyn* is an interesting one. With the greatest range of aloes found in southern

[92] Refer to Henry Hemming, *The Ingenious Mr Pyke: Inventor, Fugitive, Spy* (New York, Public Affairs, 2015).

[93] J. Cronje, 6 August 1940, File C6: J. Cronje—Aalwyn and Magawa Trees as Food for Soldiers, KMG, Group 27, Box 178, DOD.

[94] Secretary for Agriculture and Forestry, 30 November 1940, File C6: J. Cronje—Aalwyn and Magawa Trees as Food for Soldiers, KMG, Group 27, Box 178, DOD.

Africa and an illustrious history of use as objects of study and ornamentation throughout the world, botanist Illtyd Buller Pole-Evans suggested in 1919 that they could conceivably "be adopted as South Africa's national emblem". The use of "our" in the title of his article suggests a sense of proprietorship and of nationalism.[95] It seemed appropriate then for Cronje to recommend an iconic South African plant to prevent "many brave heroes [from] dying of thirst".[96]

From the use of indigenous plants to sustain troops in battle, members of the public also used the broad concepts of "ideas and suggestions" to air social or individual concerns. In his discussion of the relationship between war and social change, Arthur Marwick suggests that it may be a useful exercise to consider war not as something distinct or "alien" from society, nor is it a "catalyst" for change; war could be better understood as being an integral part of society: the notion of "society at war". War may thus react with ongoing features of society, leading to changes that may or may not be temporary, but which are evident during the conflict and in its immediate aftermath. For Marwick, there are four key changes in society that can be detected during war. The first is the "disruption" caused by war which is both negative—in terms of the devastation to life and property—but can also be liberating in allowing for the exposure to new environments. The second feature is the necessity of implementing change in institutions and other societal features to meet the needs of war and this can include shifts in industry and manufacturing. This also includes changes in the use of science and technology so as to be "more productive". The third feature relates to the "participation" of hitherto marginalized groups in wartime economy and society, especially on the home front. While this conventionally refers to the role played by women in the war effort in industry, for instance; in a country like South Africa it also applies to the opportunities available for black workers as their labour was used to compensate for the shortage of labour due to the exodus of white men and the expansion in industry. The final aspect is the psychological experience of war that may promote a sense of nationalism due to a distinction made between the nation and the enemy. This can create stronger social cohesion, evident in social welfare, for example, but

[95] I.B. Pole-Evans, "Our Aloes: Their History, Distribution and Cultivation", *The Journal of the Botanical Society of South Africa*, 1919, pp11–16.

[96] J. Cronje.

also in an agitation among the marginalized for greater political rights.[97] All of these were evident in South Africa during the Second World War and the consideration of the marginalized, in particular, is evident in some of the submissions made to the Inventions Committee.

One of the most compelling submissions was not however by a South African but by a British veteran of the First World War. File H-1 of the Quartermaster General Group contains a folded A3 scrapbook. Its cover is brightly coloured in the style of the recruiting and propaganda posters of the time and depicts a fluttering Union Jack with the French Tricolour behind it. Along with images of aeroplanes and warships is a smiling soldier in the foreground, his rifle slung over his shoulder. The scrapbook had been produced for those interested in saving "photographs and press cuttings"—an indicator of the already historical importance given to the conflict. For Albert Haskin of Newcastle-on-Tyne, however, the scrapbook was used to record his inventions. According to his accompanying letter, Haskin had served in the First World War and had subsequently been employed in the coal mines. He had then lost his job more than a decade previously and had been unable to find gainful employment since then. Upon the outbreak of the Second World War, he had attempted to enlist twelve times but had been repeatedly rejected. Unemployed, he considered his natural inclination to be that of a "draughtsman" and this was reflected in the illustrations that adorned the scrapbook—some thirty-eight in all. Each plan (or sketch) was rendered in red and blue pencil and the content was wide-ranging from a channel tunnel between England and France to new versions and different forms of armament for aeroplanes, ships and submarines. There was an electric yacht and an electric aircraft carrier—both bristling with armaments with the former holding a "collapsable airoplane" [sic] and the latter envisaged at a length of 1660 feet with a crew complement of a thousand along with 166 aircraft and ten "depht [sic] charges"[98] (Fig. 2.4).

Even a Spanish galleon, complete with sails, had been re-envisioned as being equipped with an electric "or sailing" engine and 31 "proppellers, [sic]" giving it a maximum speed of 48 knots. A "flying baby submarine" had 26 crew at sea and would then be equipped with wings and a smaller

97 Marwick, *The Deluge*, pp16, 17–20.

98 Albert Haskin, 26 October 1940, File H-1: Mr Albert Haskin, KMG, Group 27, Box 179, DOD.

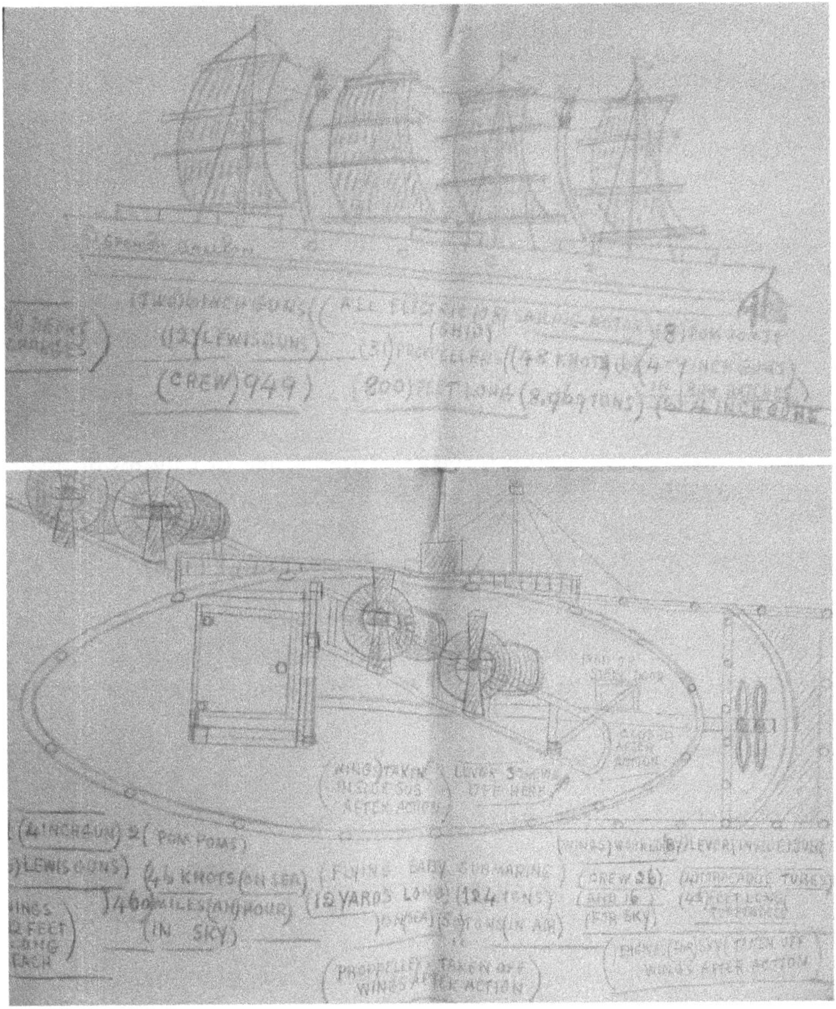

Fig. 2.4 Albert Haskin, 26 October 1940, File H-1: Mr Albert Haskin, KMG, Group 27, Box 179, DOD

crew of 16 (presumably to reduce the weight) when in flight. Each design was largely a variation of the other. While clearly indicating some thought had been put into them, the sketches had more of a science fiction rather than practical bent. Figures regarding length, speed and armament were clearly unsubstantiated and there was no indication of engineering experience. It was instead a fantastical vision of what inventions should be that enforced the idea of the individual amateur at work. Haskin's motivation for his submission—into which he had clearly put a lot of time and effort—was perhaps related to his contribution to the war effort. It was, however, also largely pecuniary. In his accompanying letter, he claimed to have submitted forty designs to the United States and was informed that these were valued at $490 which was a significant underestimation. Unemployed and unable to enlist, he was in an unenviable position and turned to the UDF for relief: "I know South Africa will help me on".[99] His hope, however, was in vain. While usually dismissive of utterly unfeasible suggestions, Ractliffe demonstrated a sense of empathy with Haskin's plight. Yet, after thanking him for his submission, Ractliffe's rejection was definite: "Your sketches have been duly examined and it is with much regret I have to inform you that there is nothing sufficiently novel in any of them to warrant further interest on our part".[100] Yet Ractliffe's choice of "novel" as grounds for rejection is itself indicative of a level of tact. There was much that was novel about Haskin's submission however his designs were entirely unviable.

A number of related ideas were also submitted by a discharged member of the UDF, Maxwell William Lempke that also aroused the interest of the Department for Social Welfare.[101] Lempke who stated his occupation as "engineer" upon his enlistment in 1940 and had been attached to the Base Workshop in Premier Mine in Johannesburg a year later,[102] had also written to Jan Smuts's Private Secretary in connection with his invention. In the letter he claimed to have been discharged from the Army without being able to complete his work and that his transfer to Cullinan physically

[99] Albert Haskin.

[100] G. Ractliffe for DGWS, 23 January 1941, File H-1: Mr Albert Haskin, KMG, Group 27, Box 179, DOD.

[101] Letter by J.H. Joubert, Secretary for Defence, "Discharged Soldier No. 227151 M.W. Lempke: Inventions", File 380: Body Armour to Protect Soldiers and Airmen (Bullet Proof Clothing)—General Suggestions, KMG, Group 27, Box 165, DOD.

[102] "Lempke, Maxwell William", Service Card, 227,151, DOD.

prevented him from exploiting the "sea matter" in Cape Town (where he resided) to develop his bulletproof jacket. He requested therefore to be re-enlisted, preferably in the Air Force at a higher rank than the one he had held (which was that of corporal). He also specifically requested being based in Cape Town, to be given funding and a "billet" to continue his work and to have his formula tested at altitude. For this prospective inventor, his stint in the Army had come at high financial cost with his savings and pay invested into research that was not developed.[103] Lempke's subsequent missive to the DGTS therefore demonstrated a sense of frustration and of desperation. The invention proposed was a bulletproof jacket and he claimed to have carried out practical experiments, determining that the "jacket lining" was capable of impeding a bullet fired from approximately 180 metres away. Further, the material used in the jacket also had the potential of being transformed into an "explosive" or "incendiary", capable of burning for up to fifty hours, and the process used in this transformation also made it potentially suitable as a "synthetic rubber". The jacket itself would weigh about 3 kilograms and could be worn over the regulation khaki uniform. Lempke claimed to be still in the experimental process and had enlisted in the UDF with the hope that he would be given the "materials" required to complete his work—which had been a financial strain on him. He bemoaned the lack of assistance which was further compounded by him being transferred "far from [his] field of research". Lempke was clearly aggrieved at the lack of acknowledgement for his "loyalty to the state" when the government had encouraged invention "to further the objects of the war for Freedom". His concluding statement was based on a combination of factors: his medical fitness resulting in his discharge, the perceived poor reception given to his proposal and the difficult financial position incurred by loss of income and an inability to continue his work:

> My experience, however, has been one fraught with bitter disappointments and heartaches, and all I asked for was to serve in a Dual Capacity at the Coast, but for some reason my project has so far fallen on deaf ears. I sacrificed every penny I had saved to do my duty to my country and now am left with my lifes [sic] ambition incompleted [sic]. Discharged "Medically Unfit" from the Army, yet I defy the Authorities to show me a

[103] Letter [unsigned], 17 April 1942, File 380: Body Armour to Protect Soldiers and Airmen (Bullet Proof Clothing)—General Suggestions, KMG, Group 27, Box 165, DOD.

> fitter man in the whole "Universe", only that my nerves are broken down through worry and the want of completing what I had set myself out to do. And who are the sufferers? None other than my Wife, Children and Self.[104]

Yet Lempke's plea would come to naught—likely impelled by desperation, there was little that was concrete in his submission and the response received by the DGTS was terse: "…in the absence of particulars concerning the suggestions put forward by you, no action with respect to the development of the materials concerned can be undertaken by this Directorate".[105]

Sergeant Melman's submission had a happier outcome. Born in January 1919 and just 22 when he enlisted, Leonard Melman listed his occupation as "Turner".[106] In 1942, Melman submitted an idea to his commanding officer for a "Synthetic Trainer" for "Air Gunners". Whereas a bombsight (such as the Norden bombsight developed and used by the United States) was an early computerized bombing system and was what Melman described as "scientific", aerial gunners had to rely on split-second decision-making in firing at often mobile targets. To do this effectively required experience. Unfortunately, experience could only be acquired in combat. Melman therefore suggested a ground-based trainer that could simulate the conditions of aerial combat, with training in "relative speed sighting or tracer sighting", allowing gunners to anticipate the movements of aircraft, adjust their targeting accordingly and increase their accuracy. For Melman, this would then result in "a minimum waste of time, expense and human lives".[107] As an instructor, Melman was perfectly placed to understand the requirements for training and how to implement them. He was given the resources he required to develop his idea and just three months after his proposal had been submitted to the

[104] Letter [unsigned], "Re: Bullet Proof Jacket, Incendiary and High Explosive", 20 May 1942, File 380: Body Armour to Protect Soldiers and Airmen (Bullet Proof Clothing)—General Suggestions, KMG, Group 27, Box 165, DOD.

[105] Brigadier General P.C. Miller, DGTS, "Re Bullet-Proof Jacket, etc.", 15 September 1942, File 380: Body Armour to Protect Soldiers and Airmen (Bullet Proof Clothing)—General Suggestions, KMG, Group 27, Box 165, DOD.

[106] "Melman, Leonard", Service Card, 208793 V, DOD.

[107] Secretary, "Ideas, Suggestions and Inventions Committee", 3 September 1942, File 416: Training of Air Gunners: Synthetic Trainers—Inventor: Air/Sgt L. Melman, KMG, Group 27, Box 167, DOD.

Committee, his Wing Commander reported that the Armament Synthetic Trainer was due to be completed under Melman's instruction within the next two weeks.[108] Once the War had ended, Miller, the secretary of the Inventions Committee endorsed Leonard Melman in a letter to the Registrar at the University of the Witwatersrand to be enrolled in a course in engineering specifically aimed at ex-soldiers.[109]

While Lempke's submission demonstrates the social nature of invention and the individual aspirations that arose from it contextualized by the war effort, his suggestions were vague and his demand for financial reward and to be given the resources to complete his research went unheeded. Melman, on the other hand epitomized the ideal inventor—a man of military experience who saw a need and devised a solution. His potential was recognized and encouraged.

In addition, while inventions could realize potential and create opportunities for the inventor, they could also reflect the prevailing divisions and exclusions in South African society. Ray Kenneth Allan penned a voluble letter addressed to Jan Smuts care of the Prime Minister's private residence in Irene. Along with a description of a potential invention—"rocket buoys" to be placed along the British coastline to deter or hinder a possible German invasion in the wake of the Battle of Britain—Allan also penned the lyrics of a song, "General Smuts' Call to the Nation". It was a stirring refrain for white South Africans to surmount the differences that divided them and to "unite the South African nation". As such it reflected in part Smuts's own concept of a white "inclusive" nation in contrast to the more exclusionary vision of Afrikaner nationalists. Part of the chorus is as follows:

> Come my sons unite and give
> Our country must remain
> Afrikaaner [sic], Britisher to live
> As one in this domain[110]

108 Wing Commander, No 44 Air School, Grahamstown, "A/Sgt L. Melman", 9 December 1942, File 416: Training of Air Gunners: Synthetic Trainers—Inventor: Air/Sgt L. Melman, KMG, Group 27, Box 167, DOD.

109 P.C. Miller, "Engineering Course: Ex Sgt L. Melman", 17 November 1945, File 416: Training of Air Gunners: Synthetic Trainers—Inventor: Air/Sgt L. Melman, KMG, Group 27, Box 167, DOD.

110 Ray Kenneth Allan, 7 January 1941, File A-1: R.K. Allan - Rocket Buoys, KMG, Group 27, Box 179, DOD.

Smuts's personal secretary replied to Allan and duly forwarded the letter to the DGWS for Ractliffe's consideration. His conclusion was that the rocket buoys were "not impracticable" but there already existed more suitable devices. With regard to the song, his comment was a clipped, "This is outside my sphere".[111] A more diplomatic response was sent to the prospective inventor and songwriter with his suggestion of the buoys exciting "considerable interest…[yet] impossible of adoption in view of existing arrangements in this direction". Of the song, there was no mention.[112]

Writing in the watershed year of 1942, Johannesburg resident Kathleen Christensen re-iterated a proposition that she had put forth a month previously in the newspaper, *The Star*: the training of women in the use of arms. She called for a repeal of any legislation prohibiting women from doing so, arguing that it should be possible for new legislation to be formulated so as "to enable the people to meet the danger with all their strength". Her recommendation—and one that she was confident many women would share—was for women to undergo "a thorough training in war-fare, exactly as the men do…[and to] learn to use tommy guns, AA guns, in fact any form of defence".[113] Interestingly enough, Christensen was considered a "suggester" rather than an "inventor", in contrast to the designation applied to those submitting similar suggestions that were applied to black men, as discussed below. Further, as her comparison to "men" undergoing combat training inevitably meant white men, it is likely that her use of "women" referred solely to white women who were the only women to form part of the auxiliary services attached to the various branches of the UDF: the Army, the Navy and the Air Force, all of which were to fall under the Women's Auxiliary Defence Corps (WADC).

As Esté Kotzé has shown, the WADC operated in similar ideological fashion in terms of appropriate gender roles to its counterparts in the United States and Britain. The auxiliary services permitted women to participate in war—an historically masculine enterprise—but in a manner that would not ultimately challenge idealized gender roles: "Women

[111] G. Ractliffe, 27 January 1941, File A-1: R.K. Allan - Rocket Buoys, KMG, Group 27, Box 179, DOD.

[112] E. Niemeyer, DGWS, 4 February 1941, File A-1: R.K. Allan - Rocket Buoys, KMG, Group 27, Box 179, DOD.

[113] Letter by Kathleen Christensen, 1 March 1942, File 394: Military Training for Women—Suggester: Kathleen Christensen, KMG, Group 27, Box 178, DOD.

cannot wholly replace men. They are by force of circumstances supplementary to men. This principle applies particularly in the military forces [...] If this principle is departed from it will completely upset our military and social structure".[114] Christensen, then, was challenging a "combat taboo"—and one which would remain in place for decades, long after black men were permitted into combat. In the South African context during the War, women could serve in anti-aircraft batteries as members of the Women's Auxiliary Army Services (WAAS) and were thus potentially on the front lines. They received extensive training, worked alongside male gunners but were ultimately prohibited from firing the guns themselves.[115] Christensen's suggestion was innovative but perhaps too much so in its violation of gender norms. As the more tentative "Suggester", she was excluded from the role of "Inventor".

Similarly, the DGWS forwarded a suggestion from A. Fawcus of Natal, described as a "Proposer".[116] Referring to himself in the third person, "Alfred Fawcus OBE" took for granted the existence of women soldiers. Alluding to a request by the government for civilians to turn in the firearms for the war effort due to a shortage, he felt it the perfect opportunity to consider the ideal weapon for women in the WAAS. They would require a "lighter weapon" such as a 12- or 16-gauge shotgun and he deliberated on whether its efficacy would be increased if the shotgun were "sawn off"—making it lighter but only really effective at closer quarters.[117] Fawcus was thanked for his proposal—but he had clearly jumped the gun and the response by the DGTS was that his suggestion "would be borne in mind when occasion arises".[118]

Another group forming part of the auxiliary services—who were also excluded from combat—were black men. From as early as the South African War at the turn of the century, there was opposition to the taking

[114] Quoted in Esté Mari Kotzé, "More Than Just Pretty Girls in Uniform: A Historical Study of Women's Military Roles during World War II, 1939–1945" (Stellenbosch University, Unpublished PhD Dissertation, 2021), p42.

[115] Kotzé, "More Than Just Pretty Girls in Uniform", p126.

[116] R.R. McGregor, DGWS, 30 July 1942, File DTS/INV 413: Arms for WAAS—Proposer: A Fawcus, Dunstanburgh, Eston, Natal, KMG, Group 27, Box 166, DOD.

[117] Alfred Fawcus, 12 June 1942, File DTS/INV 413: Arms for WAAS—Proposer: A Fawcus, Dunstanburgh, Eston, Natal, KMG, Group 27, Box 166, DOD.

[118] P.C. Miller, DGTS, 10 August 1942, File DTS/INV 413: Arms for WAAS—Proposer: A Fawcus, Dunstanburgh, Eston, Natal, KMG, Group 27, Box 166, DOD.

up of modern weaponry by black—and specifically indigenous African—men, especially if these weapons could potentially be used against whites. With the outbreak of the Second World War and the desire on the part of Afrikaner nationalists to remain neutral along with fears of arming black men—which could potentially present an internal security problem—the black men falling under the NEAS were officially designated as auxiliary non-combatants. Their weapons of war were assegais (or spears) with which they trained and were armed while on sentry duty.

W.H. Blower however, submitted an idea for the use of black men in combat based on his observation of "native youths" using "kerries"[119] which were aimed with unerring accuracy to hunt birds in flight as well as animals. Blower went on to propose the creation of a special unit of these men who would employ expressly adapted "kerries" that were armed with "a grenade, small bomb, or bottle of petrol" against an invading force attempting a beach landing or against tanks and other motorized vehicles. Blower also emphasized the perceived attributes of these men: "They are experts at stealthy approach, taking cover, cross country running". He also attempted to allay the concerns of the arming of black men by stressing that this unit would only be used in a worst-case scenario, "when actually needed". Blower's "invention" then challenged the status quo—but did so within existing boundaries yet also made use of an indigenous weapon that could be adapted as a tool for modern war but only with western ingenuity and originality, with Blower cited as the "inventor". But even more significant was the response by the Inventions Committee: a handwritten comment stated, "should receive further consideration by defence authorities" which was endorsed by another comment, "I certainly agree". The recommendation was "that the idea is well worth investigating" and it would subsequently be forwarded to the Non-European Army Services for consideration.[120]

[119] A "knobkerrie" is a wooden stick with a knob at the end.

[120] Letter by W.H. Blower, 10 March 1942, File 393: Training of Natives on Military Lines—Inventor: W.H. Blower, KMG, Group 27, Box 178, DOD. Blower's recommendation is reminiscent of a proposal put forth by Charles Edward Callwell, Director of Military Operations and Intelligence at the War Office during the First World War. Callwell was approached by a Captain Martin who suggested the use of "Basuthos, Zulus and Swazis" as prospective "bomb-throwers". While these ethnic groups had had little experience in the use of modern weaponry, they were nevertheless skilled in terms of their aim and accuracy where "A boy in a kraal out there can knock a bird over on the wing". - Letter from Charles Edward Callwell, Director of Military Operations and Intelligence at

Just five days later, another letter was received that argued unequivocally for the arming of the black men, while acknowledging that "a certain section of South Africans will not agree". For the writer, the potential invasion would come via paratroopers and South Africa should prepare by training "thousands of natives" to be dispatched to potential drop zones. They would be under the command of white officers and be armed with "rifles, assegais, bows and arrows". As with Blower, it was a compromise between the "traditional" and the "modern". While the writer disarmingly concluded by conceding their lack of knowledge, thus rendering their "suggestion…useless", the reasoning behind the statement was both the defence of a homeland in peril but also to "give the native a chance to do his bit for his country".[121]

Both submissions appeared as the policy of dilution was being implemented in the UDF. Due to a recruiting shortage that would later be compounded by the surrender at Tobruk, black men from NEAS units were incorporated as non-combatants in previously white units within South Africa as well as in the North African theatre. While unarmed, they were subjected to similar dangers of coming under enemy fire as their white counterparts as they went about their duties as drivers, stretcher-bearers and mechanics. Dilution highlighted both the unequal treatment afforded black non-combatants but also opened the way for possibly greater co-operation and camaraderie. 1942 also saw the Communist Party of South Africa intensify their efforts to provide greater equality in the workplace for black workers and to arm black men via their "Defend South Africa" campaign. "Loosen our hands to defend South Africa" was the refrain in a "Non-European People's Manifesto".[122]

The exclusion of black men from combat reflected their exclusion from citizenship—but also from invention. Drawing upon Adas, Edgerton argues that, whereas technology took on both global and national implications, this paled in comparison to the racial distinctions made within

the War Office to L.H. Robertson, 11 November 1915, LH Robertson Collection, 7/2/36, Liddell Hart Centre for Military Archives, King's College Archives, London. My thanks to Tilman Dedering for providing this reference.

[121] Letter by S.N. Buiers, 15 March 1942, File 393/1: Training of Natives for Defence of South Africa—Inventor: S.N. Buiers, KMG, Group 27, Box 178, DOD.

[122] Suryakanthie Chetty, "'Subjects or Citizens?': Black South Africans and the Dilemma of the Second World War", *Journal of Natal and Zulu History*, 53–79, Vol 30, No 1, 2012, pp75–76.

states and within the context of empire. "A sense of scientific and technological superiority" was inherent in Western "civilization" and a mark of distinction between white men and those who fell under their governance, and technology was utilized as part of the "civilizing mission" and of conquest.[123] South Africa was no exception. Dubow has demonstrated that the history of science and scientific institutions in South Africa since the inception of colonization was one that was intertwined with a British, settler and later exclusionary nationalist identity.[124] Beginning with the mineral discoveries in the nineteenth century, the South African economy was underpinned by the control of black labour and the stifling of the emerging black middle class. As such, education opportunities—and higher learning—were limited for the country's black majority. These inequalities were reflected in the submissions received by the Ideas, Suggestions and Inventions Committee. Suggestions were not categorized according to existing racial categories and all members of the public were encouraged to submit their ideas, but the bulk of submissions appear to be by white men—some of whom had either engineering or previous military experience—most of which were written in English, although there were some submitted in Afrikaans as well. They largely appeared to mirror the key figures in the War itself, and the country's ideal citizens—white, male combatants. In April 1940, a brief excerpt appeared in the UDF magazine, *The Nongqai*, detailing a "native" invention. The invention—an adapted rifle—had been discovered by police during a search "for evidence in a cattle poisoning case". Described as the "Rifle" (in quotation marks), the inventor had taken a discarded rifle barrel, cleaned it, added "brass tubing" and created a firing mechanism from an "airgun spring" and wire. Displaying "wonderful…ingenuity", the inventor was charged with having an unlicensed firearm and the rifle was confiscated and sent to the Police Museum in Pretoria to be displayed as an item of curiosity. In contrast to the image of South African inventions as symbolic of a citizen's contribution to the war effort, this "native" invention was suggestive of criminality and with the article titled "Something Like

[123] Edgerton, *The Shock of the Old*, p132.

[124] Refer to Saul Dubow, *A Commonwealth of Knowledge: Science, Sensibility and White South Africa 1820–2000* (New York, Oxford University Press, 2007).

Hitler's Secret Weapon!", there was the implication not just of absurdity but of a subversion of the war effort and the status quo.[125]

Similarly, the concept of "inventiveness" itself was ideologically loaded. Seabury Colum Gilfillan graduated from Columbia University with a doctorate in sociology. A devotee of H.G. Wells, he is considered a seminal figure in "future studies", predicting the invention and future use of technology.[126] In 1930, he published an article in the *Geographical Review* containing a table with the list of patents for each country in 1925. In pole position was Switzerland with 167 patents; the United Kingdom was ranked 8th; the United States 12th; the USSR at 28 with South Africa in 19th position with 3.7 patents. While acknowledging that patents were an inaccurate reflection of invention, Gilfillan's article is less about his statistical analysis in relation to the inventive strength of countries than it is about who was excluded from this analysis. At the outset, he considers it "inadvisable" to include "the coloured populations of the United States and the British dominions". While these regions are numerically strong—in contrast to Switzerland, for instance—"these people do not figure in invention". In his conclusion then, Gilfillan deemed "race to be a strong factor" accounting for inventiveness. Invention was inevitably linked to "modern industrial civilization", evident in Switzerland and in the United States. Inventions in less industrialized British India were the product of "the handful of European residents". Switzerland then, exemplified this complex mix of racial "purity" with industrialization and the production of goods that accounted for its pre-eminent position.[127] Similarly, albeit writing from a less statistical perspective some three decades after the end of the Second World War, South African reconstructive surgeon Jack Penn would associate inventiveness with a racial hierarchy. It was the ability to have "original thought" and to apply this, to be "inventive" and to have "mental flexibility" that marked those who had reached a higher stage of development. Those "groups" who had only the ability

[125] "Something Like Hitler's Secret Weapon!", *The Nongqai*, April 1940, p459, DOD.

[126] Refer to Matthieu Ballandonne, "The History of Futures Studies: A Note on Gilfillan's Early Work", *Technological Forecasting and Social Change*, Vol 157, August 2020.

[127] S.C. Gilfillan, "Inventiveness by Nation: A Note on Statistical Treatment", *Geographical* Review, Vol 20, No 2, 1930, pp301–304.

to "assimilate facts" but lacked "originality" still needed a greater degree of "[evolution]".[128]

Yet this may simplify the multifaceted ways in which ideas of invention took shape in the South African context, such as the suggested use of aloe—used in local healing—to slake the thirst of South African troops in the North African theatre as discussed above, as well as the special unit armed with "kerries". Therefore, despite the historic distinctions and disparities in South African society, the submissions—albeit a minority of them—allowed for a questioning of existing social and ideological structures on the part of members of the public and it was part of the mandate of Committee to give these views due consideration. It is not just the suggestions made by the public but the due diligence and actions taken by the Committee—even if these suggestions did not fulfil the criteria for invention and did not lead to significant change—that suggest that this wartime Committee reflected the possibilities of social change wrought by the upheaval of war where ordinary South Africans were given the opportunity and the forum to envision changes in society.

In May 1944, the Committee was reconstituted as the "Inventions Sub-Committee" and would continue to hold its regular meeting at Impala House, headquarters of the Technical Services Directorate, in Pretoria.[129] In its various incarnations since 1940, the Committee had considered submissions for prosthetic and medical devices,[130] suggestions highlighting socio-economic conditions in wartime South Africa and an overwhelming array of proposed inventions of offensive and defensive technology. These included the mundane such as wind-screen wipers, batteries, engines and petrol-saving devices; the ambitious in the form of jet propulsion, an aircraft carrier hangar and a "submerged power plant;" and the necessary: the means to clear minefields,[131] mines and bombs, rockets, flame throwers and decoding machines.[132] Submissions fell along a broad spectrum from a simple speculative paragraph to a well-conceived

[128] Jack Penn, *The Right to Look Human: An Autobiography* (Johannesburg, Hugh Keartland Publishers, 1976), p251.

[129] DGTS, 17 May 1942, File 320/1: Minutes of Meeting etc. Main Inventions Committee—Mr W.E. John's Personal Copies, KMG, Group 27, Box 178, DOD.

[130] The Brenthurst Splint and the acrylic eye are discussed in detail in Chapter Six.

[131] The flail device and the dispute over invention is the focus of Chapter Four.

[132] Refer to File 320/1: Minutes of Meeting etc. Main Inventions Committee—Mr W.E. John's Personal Copies, KMG, Group 27, Box 178, DOD.

plan, replete with diagrams and calculations. They were the work of the amateur and the expert. Some were tested and put into development, a few forwarded to Britain or the United States for consideration, many dismissed—but all were acknowledged.

In a speech at the opening of the British Association for the Advancement of Science in September 1941 (that was partially reproduced in the South African *Rand Daily Mail*), American ambassador to Britain, John Winant stated:

> The Nazi power has, with relentless cruelty and efficiency, degraded the dignity of the individual and attempted to abolish the rights of man. Nazism has stolen and run amok with great inventions, and is using them to enslave the human spirit.[133]

For the Allies then—and for South Africans—the submission and consideration of inventions were part of the democratic process, of allowing individuals to play a role in the war effort, of acknowledging individual ingenuity and invention amidst the dehumanization of modern warfare and the ideology of totalitarianism. Inventions also reflected the ways as proposed by Marwick, in which war presented opportunities for social change. In South Africa, the submissions received by the Ideas, Suggestions and Inventions Committee were indicative of the fears and anxieties, the contradictions, limitations and simultaneously, the possibilities of a society divided—that was nevertheless hovering on the brink of change through its participation in the War.

[133] "Churchill Sends Message to Scientists", *Rand Daily Mail*, 27 September 1941.

CHAPTER 3

"Strange and Wondrous Things": Military Technology on Show

The Second South African War (1899–1902) pitted two very different military systems against each other in what was to be the most extensive war fought on the African continent in the nineteenth century. At the beginning of the century, the British Army was led by an elite class of officers considered to be "amateurs" and "gentlemen". The shortcomings of the Army were exposed in conflicts during the century, including the first war with the Boers in 1881. Towards the latter part of the century then, the British Army underwent a process of reform that emphasized professionalization and, on the part of working-class recruits, discipline; instilling of a sense of duty; and training them to both "obey" and to "command", that would transform them into respectable members of society. There remained however, resistance to professionalization, as lamented by figures such as Herbert Kitchener, commander-in-chief of British forces during the latter phase of the South African War. Officers were usually drawn from the ranks of the elite and the public school educated, and part of a system of patronage.[1] Some similarities existed

[1] Refer to Gwyn Harries-Jenkins, "The Development of Professionalism in the Victorian Army", *Armed Forces and Society*, Vol 1, No 4, 1975, pp472–489 and David French, *Military Identities: The Regimental System, the British Army and the British People c 1870–2000* (Oxford, Oxford University Press, 2005), p26.

S. Chetty, *Science and Technology in South Africa, 1939-1946*, Palgrave Studies in the History of Science and Technology,
https://doi.org/10.1007/978-3-032-05084-7_3

with the Boer commando system. As Ian van der Waag shows, while officers were elected this, too, was part of a system of patronage leading to often ineffective leadership. The commando system had its origins in the early days of the Dutch settler presence at the Cape where, in the absence of a standing army, able-bodied men were liable for military service. Optimistically referred to as a group motivated by the spirit of volunteerism and teamwork, they were independent-minded, often poorly trained and chafed under military discipline.[2]

The amalgamation of the two former Boer Republics and British colonies into the Union of South Africa in 1910 required the creation of a modern military system for the twentieth century. The Union Defence Force (UDF) came into existence with the South African Defence Act of 1912. Initially expected to be a compromise akin to the formation of the Union itself, with the combination of both British and Boer military systems, Minister of Defence Jan Smuts decided instead on a professional military system that drew inspiration from the western model, in contrast to the Boer commando system.[3] Drawing criticism from the Afrikaner sector for its departure from the Boer martial tradition, the UDF exemplified the split in white society that would come to the fore with the outbreak of the Second World War. Simultaneously and contradictorily, its modern bent and accentuation of military technology, innovation and industrialized war would also be used to rally support for the war among an often hostile populace.

Military strength has been displayed to the public since the very beginnings of an organized military system. Currently most associated with autocratic regimes, evidence of military parades and the triumphant return from battle of kings and emperors are apparent in ancient Babylon and Rome. In the modern era, the precision and discipline associated with the military parade has its origins in Prussia and reached its zenith with the rallies in Nazi Germany which were meant to awe, to overwhelm and to assert the overwhelming power of the state.[4] As such, they are not often associated with democracies with its checks on rampant militarism. However, in times of war where there was a need for civilian

[2] Ian van der Waag, *A Military History of Modern South Africa* (Jeppestown, Jonathan Ball Publishers, 2015), pp19–21.

[3] Van der Waag, *A Military History of Modern South Africa*, pp73–74.

[4] Ishaan Tharoor, "Military Parades", *Time*, 1 October 2009, https://time.com/archive/6948154/military-parades/, accessed 10 June 2024.

support—a need not necessarily acknowledged in totalitarian states—military parades demonstrated the confidence associated with the resources needed to wage a successful war. In the United States in April 1942, just five months after the Americans entered the War, a massive parade comprising thirty thousand men and women drawn from all branches of the military was held in New York. For the Americans this was seen as the "first big military display of the war"—and it was a display designed for the home front.[5]

With perhaps less unabashed celebratory overtones, is the military museum exhibition. Taking cognizance of civilian fascination with the weaponry, uniforms and other accoutrements associated with military service, the military museum began in the nineteenth century—and is therefore contextualized by the rise of the nation-state. The late nineteenth century also saw the display of military weapons and equipment at the international exhibitions in Europe and the United States. As Hacker and Vining argue, these tended to be "more evocative than substantive", focusing on the weapons of war, equipment and uniforms that did not always reflect the comprehensive nature of the war effort in both the civilian and military spheres.[6]

It is therefore within this patriotic fervour—evoked by the outbreak of war—the tradition of the exposition and the ceremony of military display that the UDF attempts to showcase military displays to the public can be located. In South Africa, however, this was complicated by the competing nationalisms and divided loyalties exacerbated by the War. This chapter addresses then the Rand Show with a focus on the exhibition held in 1940, the Air and Steel Commandos, the War Train and the Liberty Cavalcade, to consider the ways in which military technology was packaged for a fragmented country.

[5] Maris Fessenden, "The History of Military Parades in the US", *The Smithsonian Magazine*, 7 February 2018, https://www.smithsonianmag.com/smart-news/how-military-parades-us-have-changed-180968102/, accessed 10 June 2024.

[6] Barton C. Hacker and Margaret Vining, "Military Museums and Social History", *Does War Belong in Museums?: The Representation of Violence in Exhibitions*, Wolfgang Muchitsch (ed.), (Bielefeld, Transcript Verlag, 2013, pp41, 47.

The Rand Show

The South African National Defence Force (SANDF) was expected to maintain a strong presence at the Rand Show to be held at the Johannesburg Expo Centre over the Easter weekend in 2024. All four branches of the military (the Army, the Navy, the Air Force and the Military Health Service) would be exhibiting their technology. For the Army, these were Hornet reconnaissance vehicles—a lightweight armoured car—in addition to boats, weapons and survival gear. The Air Force would have a Rooivalk helicopter, radar and an ARGOS surveillance system usually fitted to planes or helicopters and used for reconnaissance. The Navy planned to display the weapons associated with their branch of the service. Due to the various economic challenges faced by South Africa and in light of corresponding budgetary constraints, the military display at the Rand Show was envisaged as being more modest than those in preceding years which historically included far more ambitious military technology in terms of fighter aircraft for instance.[7] It was also in contrast to the elaborate military exhibition at the Rand Show during the Second World War. Initially, used to showcase cattle from the late nineteenth century, the Rand Show took on a greater industrial and military bent over the course of the twentieth century. The last show held during the War was in 1940—after which the showgrounds were repurposed for military use until 1946.[8]

At a meeting of the organizing committee for the Defence exhibit, Colonel F.H. Theron stated that the objective of the exhibit "was to show the public how industry was being organized to produce munitions…for the defence of the country". For the Director General of War Supplies, H.J. van der Bijl, just as important as the industrialization for war (for which he was directly responsible) was South Africa's tradition of war and his emphasis was on "a historical review of armament manufacture in South Africa" where the early "Sanna" had been constructed

[7] Guy Martin, "SANDF to be out in force at Rand Show", 22 March 2024, https://www.defenceweb.co.za/sa-defence/sa-defence-sa-defence/sandf-to-be-out-in-force-at-rand-show/, accessed 24 March 2024.

[8] Marc Latilla, *Johannesburg: Then and Now*, https://johannesburg1912.com/2015/04/29/history-of-braamfontein-pt-4-rand-show-wits-nunnery-brewery/, accessed 6 February 2024.

from "wagon tyres".[9] As David Edgerton points out in *The Shock of the Old*, while technology has often been associated with the innovative and the new, older technology has been incorporated into newer technologies. In contrast to the teleological perception of technological progress, "…technologies do not only appear, they also disappear and reappear, and mix and match across the centuries".[10] Evident at the Rand Show—and throughout the many exhibitions and tours undertaken by the UDF—was this mix of the old and the new with a particular emphasis on the ways in which "tradition" was portrayed in order to create a sense of unity and patriotism—and to increase support for a very modern war.

By the beginning of March, preparations were underway for the UDF exhibit at the Rand Show. From the outset and just six months after South Africa joined the War, the exhibition was deemed to be significant with the Minister of Defence "[desirous] that the greatest publicity be given to the Defence Force Exhibits and activities at the show".[11] Colonel F.H. Theron, a member of the organizing committee for the exhibition, considered the various military equipment "most likely to be of general interest" for the South African public. These included aircraft such as "a modern high speed bomber" as well as the Wapiti, Tutor and Hartbees which had been produced in South Africa, a naval display including mine sweeper and a depth charge, various transport vehicles, arms and ammunition and the "historic weapons made in this country", including the "Griepjie" used by the "Voortrekkers" against the Zulu in 1838, a "Ras gun used in 1881" during the First South African War and the "Long Cecil" gun used during the Siege of Kimberley during the Second South African War. For Theron, there were four key aims of the exhibition: to demonstrate to the public the "complex…processes" of the production and maintenance of military equipment and the importance of industry to the modern war effort; the concomitant importance of the growth of

[9] Director General of War Supplies, "Defence Exhibit at Rand Show: Notes of a Meeting of the Committee formed to Organise the Above Exhibit", 5 February 1940, Ref DWS1045, Vol 146: "Defence Exhibit Rand Show", DGS, National Archives of South Africa (hereafter SANA).

[10] David Edgerton, *The Shock of the Old: Technology and Global History Since 1900* (London, Profile Books Ltd, 2008), pxii.

[11] DCGS, "Military Camp at the Agricultural Show Ground, Milner Park, Johannesburg", 1 March 1940, File A.G. 211/4/2: Witwatersrand Agricultural Shows, Adjutant General 3, Box 66, DOD.

industry; the recruitment of skilled labour to facilitate that growth; and the use of "basic training" initiated under Van der Bijl to address the shortfall in skilled labour. For this dramatic shift to industrialization then, the exhibit was to garner "the support of every section of the people".[12]

The UDF exhibition was to combine a display of both "war supplies" and the more military aspects of the War. The men placed in charge of the latter were Lieutenant-Colonel Schoon, Captain Simpson and Lieutenant Emery. They were also enabled to "co-opt additional members" as needed to ensure that "the Exhibition is a credit to the Military Forces".[13] The exhibit was not simply to make the South African public aware of the military industrial might of the UDF but to encourage recruitment, and Schoon had two recruiting offices set up—one where men could enlist for the various military branches of the UDF and the other for the labour required to produce the trappings of modern warfare.[14] And these accessories were extensive.

There was to be a fully-fitted armoured car with arrangements made with ISCOR (the Iron and Steel Corporation) for the armoured plating; a "travelling workshop", various ambulances and a "travelling laboratory". In terms of guns and munitions, preparations were made to provide a Howitzer, six 10-pound mortar bombs along with metal castings and fuses, one each of the 500-, 250- and 120-pound aircraft bombs, telescopic gun sights and even steel helmets "in various stages" of production from the "blanks" to "pressed" stages with a "complete helmet, if available". The last was an indication that, although the exhibit was designed to showcase the country's wartime industrial production, just five months into the war, that production had yet to come fully into effect. In the case of "light carriers", instructions were issued to look at the "final drawings" for the construction as well as when tenders were to be submitted. In the interim, twelve carriers composed of papier mâché would be made for the exhibition. The situation was similar for Army Transport carts still in the "amended drawings" stage and, should a completed cart not be available by the time of the exhibition, an older version would be

[12] Draft letter by Colonel Theron, Ref DWS1045, Vol 146: "Defence Exhibit Rand Show", DGS, SANA.

[13] DCGS, "Military Exhibit – Rand Show, 1940", 16 February 1940, File A.G. 211/4/2: Witwatersrand Agricultural Shows, Adjutant General 3, Box 66, DOD.

[14] Director General of War Supplies, "Defence Exhibit: Meeting of Committee", 19 February 1940, Ref DWS1045, Vol 146: "Defence Exhibit Rand Show", DGS, SANA.

used. The "trailer ambulance" was still being constructed and plans were made for the prototype to be exhibited. The "bulk production" of signal pistols was as yet not underway but the six that had been assembled had been forwarded to the UDF.[15] The Howitzer—which would arouse considerable interest at the exhibition—was a South African success story. Described as "the first and only Howitzer built in South Africa" and as a "purely South African achievement", the 3.7-inch gun had been manufactured by Selby Engineers (Pty) Ltd under the guidance of T.F. Atkinson who had previously been employed by Vickers-Armstrong, the British armaments engineering and manufacturing company. The gun had taken eight months to construct and this "sample" was readied for display at the exhibition.[16] The impression given to the public then was of capacity that had yet to be realized.

For Prime Minister and Commander-in-Chief of the UDF, Jan Smuts, key to the exhibit was the "propaganda value", particularly as it related to "military organization". Even if the exhibition came at the risk of hampering the training of soldiers for the war itself, the focus and aim of the exhibition was on gaining public support and raising awareness. As such, Smuts was desirous of the enacting of "military displays in the arena" and the Defence Exhibit was to be held in the Empire Hall (site of the Empire Exhibition in 1936 to commemorate the 50th anniversary of the establishment of Johannesburg) so that it "[would] monopolise a great deal of public attention".[17] While an initial estimate of £1,500 for the staging of the exhibition was considered to be steep, it was nevertheless accepted as "on no account [should the] effectiveness of Exhibit be impaired in order to save on estimate".[18]

Staffed by technicians that were the vanguard of the Central Organization for Technical Training, the exhibition consisted of displays of the weapons and equipment associated with war, both historically and

[15] "Preliminary Arrangements in Conjunction with Proposed Exhibit at the Witwatersrand Agricultural Show: 19th to 25th March 1940", 19 February 1940, Ref DWS1045, Vol 146: "Defence Exhibit Rand Show", DGS, SANA.

[16] Letter by Acting News Editor, *The Star* to Colonel F.H. Theron, Press Release, 14 March 1940, Ref DWS1045, Vol 146: "Defence Exhibit Rand Show", DGS, SANA.

[17] Letter to G.R.D. Harding, Escom House, "Rand Show: Defence Exhibit", 28 February 1940, Ref DWS1045, Vol 146: "Defence Exhibit Rand Show", DGS, SANA.

[18] Letter by A.C. McColm to Colonel F.H. Theron, 4 March 1940, Ref DWS1045, Vol 146: "Defence Exhibit Rand Show", DGS, SANA.

for the contemporary conflict, with the aim of increasing recruitment by demonstrating South African self-sufficiency in armament production. Included in the display was the cannon used against the Zulu at the Battle of Blood River (Ncome) in 1838, a significant moment in the story of settler conquest and Afrikaner nationalism. Another cannon mounted on wagon wheels (likely "the Sanna" referred to by Schoon) had been employed by the Boers against the British during the South African War. Dominating the exhibition, however, were the armaments produced for the Second World War: Wapiti, Tutor and Hartbees aeroplanes that, sans engine and undercarriage, were manufactured in South Africa; the Link Trainer used to train pilots along with an early version of "automatic pilot"; a reconstructed hangar manufactured from South African steel; wooden air screws made of mahogany. The process by which the Royal Mint produced munitions was highlighted as was the role of women of the South African Women's Aviation Association in packing parachutes. For the medically inclined, the South African Medical Corps had set up a fully functional field hospital. The field hospital was also conveniently able to provide physical examinations of potential recruits as a prelude to their enlistment.[19] Almost 125,000 visitors attended the exhibition which would be described as "stupendous" by Jan Smuts.[20] For visiting dignitary General Sir Archibald Wavell who had served in the Second South African War and was the Commanding Officer in the Middle Eastern theatre in 1940, the exhibition was a "revelation", indicative of the country's long and successful path to modernization:

> [The exhibit] speaks very highly indeed of the organizing ability and manufacturing skill of South Africa…Indeed the whole Show has been a revelation to me. My recollections are of South Africa in the days of the South African war, and to-day I was delighted to see the beautiful cattle that are being bred, and the wonderful agricultural and industrial progress which is being made by this country.[21]

[19] "Union's Growing 'Arms Factory' to be on View at Rand Easter Show", *Rand Daily Mail*, 8 March 1940.

[20] Latilla, *Johannesburg: Then and Now* and Letter by Director-General of War Supplies to Commander J.B. Bullock, 27 March 1940, Ref DWS1045, Vol 146: "Defence Exhibit Rand Show", DGS, SANA.

[21] "Defence Exhibit 'Stupendous' Says General Smuts", [Newspaper article], 21 March 1940, Ref DWS1045, Vol 146: "Defence Exhibit Rand Show", DGS, SANA.

Despite the many political and ideological changes over the course of the twentieth century that saw the formation of the UDF, its transformation to the South African Defence Force (SADF) and the current South African National Defence Force (SANDF), a constant has been the attempt by the national defence force to awe the public through its military technology and ability to wage industrialized and modern war. Its annual displays at the Rand Show have fostered a sense of nationalism and patriotism, despite the division and exclusion that has characterized South African society. Like the defence force itself, military displays are a reflection of the tensions and aspirations of the South African state.

While the first and last Rand Show for the duration of the War was held in 1940, this by no means marked the only attempt by the UDF to showcase South African military strength to the general populace. As industrial wartime production ramped up, displays became increasingly elaborate in their attempt to shock, awe and perhaps overwhelm the South African populace.

The Steel Commando

The Steel Commando was an early example of the UDF "roadshow" that would travel throughout the country and serve a two-fold purpose—increase recruitment and support for the war effort and display South African industrialization for the War. The appellation "Steel Commando" is a significant one. "Commando", on the one hand, evoked a settler and particularly Boer past that was only enhanced by the description of the Commando's tour as a "trek"—terminology that Smuts would adopt for emotive effect. At the same time, "steel" was intertwined with South Africa's modernization project. The Iron and Steel Corporation (ISCOR) was established in 1928 and was under the leadership of Hendrik Johannes van der Bijl. In 1919, Van der Bijl—then working for the American Telephone and Telegraph Company—penned a paper "Scientific Research and Industrial Development", advocating the growth of secondary industry in South Africa. His work brought him to the attention of Jan Smuts and Van der Bijl returned to South Africa to advise on "industrial development". He was instrumental in the formation of the Electricity Supply Commission (Eskom) in 1922 and Iscor six years later. For Van der Bijl, the state had an integral role to play in industrial development and both Escom and Iscor were parastatals. The relationship between the state and industrialization was strengthened with the outbreak of the War when Smuts appointed Van der Bijl the Director General of War Supplies. Along with Iscor manager, Hendrik J. van Eck, Van der Bijl was responsible for the industrial expansion necessitated by the War[22]—resulting in the material on display during the tour of the Steel Commando.

The Commando left Roberts Heights or Voortrekkerhoogte on the morning of 1 November 1940 on a journey of almost five thousand kilometres. It comprised armoured cars and mobile field artillery with six hundred members of the UDF, accompanied by the Permanent Force Band and Bugle Band comprising eighty musicians. Travelling through the East Rand, it stopped at the towns of Heidelberg, Standerton and Balfour on its first day. Leaflets had been dropped from the air two days prior over these towns stating simply, "The Army is coming". Leaflets were also posted to residents, and advertisements appearing in the press raised awareness of the Commando's imminent arrival. The publicity associated with the Commando was designed to draw in the crowds, evoking a sense of nationalism and awareness of the "striking might" of the UDF and its preparedness for the demands of war: "It is an army ready to compete on equal terms with any European foe".[23] Prior to its departure, Issie Smuts handed to Director of Recruiting G.C.G. Werdmuller a ceremonial scroll that would be inscribed with the names of those who were moved to enlist.[24]

A.V. Weinerlein was a young boy who had been enrolled in the English-medium Heidelberg Public School in the town of Heidelberg, just fifty kilometres away from the metropolis of Johannesburg.

[22] C. Plug, "Van der Bijl, Dr Hendrik Johannes", *S2A3 Biographical Database of Southern African Science*, https://www.s2a3.org.za/bio/Biograph_final.php?serial=2909, accessed 6 December 2024, and Willian Beinart and Saul Dubow, *The Scientific Imagination in South Africa: 1700 to the Present* (Cambridge University Press, 2021), p245.

[23] "Great Army Column to Move Tomorrow", *Rand Daily Mail*, 31 October 1940.

[24] "Recruits' Scroll", *Rand Daily Mail*, 2 November 1940.

In a town epitomizing the fractious nature of white South Africa and the ambivalence—if not outright hostility—to the war effort,[25] Weinerlein recalled the boyish joy of the arrival of the Steel Commando, giving children the opportunity to "[clamber] over the armoured cars, guns, troop carriers and other military hardware..." The soldiers attached to the Steel Commando demonstrated weaponry and attempted to recruit volunteers to enlist, with some success among those already in favour of South Africa's role in the War.[26]

Bloemfontein closed its schools and businesses allowed their workers time off to view the spectacle of the Steel Commando as it wound its way through the capital city of the Orange Free State a week into its tour. The convoy was officially welcomed by the mayor and the Special Service Battallion performed the Retreat ceremony in full dress uniform at sunset. As the convoy set up for the night, officers were treated to a dinner by Mayor C.J. Sutton and the band entertained the crowd followed by speeches by the mayor and Werdmuller.[27] While this was a scenario that would be repeated throughout the Union, it took on added significance in Bloemfontein. With South Africa's entry into the War just over a year earlier and Smuts's rise to power, previous prime minister J.B.M. Hertzog joined with Afrikaner nationalist D.F. Malan to form the Reunited National Party. But the two had different visions of Afrikaner nationalism—Hertzog may have desired South African neutrality during the War but Malan's supporters ardently supported a British (and, by association, South African) defeat. In early November 1940 in Bloemfontein, Hertzog was ultimately ousted from the National Party, becoming a

[25] A small town composed of both English-speakers and Boers, Heidelberg exemplified the tension and divided loyalties incurred by the South African War. Neighbours and friends found themselves in conflict with each other and this culminated when a Boer court martial found two Boers – who had joined the British Heidelberg Volunteers – guilty of treason. The men were executed by firing squad. Please refer to Robin W. Smith, "The Story of Two 'Joiners': The capture, trial, sentencing and execution of Pieter Bouwer and Adolf van Emmenes", *Military History Journal*, Vol 15, No 5, 2012, http://samilitaryhistory.org/vol155rs.html, accessed 18 November 2024.

[26] A.V. Weinerlein, "War Reminiscences: War experiences of a youngster in South Africa, 1939–1945", *Military History Journal*, Vol 2, No 2, 2001, http://samilitaryhistory.org/vol122aw.html, accessed 8 June 2024.

[27] "Bloemfontein Welcomes Convoy", *Rand Daily Mail*, 7 November 1940.

nonentity in South African politics until his death two years later.[28] A wounded man, Hertzog would describe the National Party as a "curse to the nation".[29] His sentiment was echoed by Smuts who refused to be cowed by the Nationalists and by Nazi sympathizers such as former Minister of Defence, Oswald Pirow. For Smuts, anything less than participation in the War and support of the Allies was "to be led on the road of dishonour". In a speech that emphasized unity, he also spoke of the passing through Bloemfontein of the Steel Commando, just as Hertzog suffered his humiliating defeat and, for the South African Prime Minister, the Nationalists with their symbol of the "ox-wagon" were emblematic of the past; the Steel Commando was "[symbolic] of the future of South Africa".[30]

And it was a future predicated on leadership in Africa—and perhaps the country's own imperial ambitions. As would be the case with the subsequent War Train tour, the demonstration of South African industrial and martial strength was a beacon of civilization on the African continent and the UDF—as exemplified by the Steel Commando—would "[open] up the [Dark Continent], not from the North, but from the South".[31]

Just as newspapers advertised the arrival of the Steel Commando in the towns throughout the Union, its triumphant return on the 10th of December was marked by advertisements and posters in the local press. One advertisement appearing in the *Rand Daily Mail* was captioned "We're not reading history…we're making it!" It depicted various types of servicemen giving a brief description of their jobs in the UDF. For the anti-aircraft gunner, "Shooting birds on the wing is a scientific business"; while the sapper lauded his training, "I couldn't build a henhouse before the war but now you should watch me run up bridges"; according to the aircraftsman, "This is THE wartime job for any chap with mechanical knowledge"; and for the medical orderly, "I've learnt a lot that's

[28] Andre Wessels, "South Africa and World War II: the decisive first two years on the home front (September 1939 - September 1941)", *Southern Journal for Contemporary History,* Vol 24, No 2, 1–24, 1999, p4.

[29] "Herenigde Party will Become S.A. Curse", *Rand Daily Mail*, 11 November 1940.

[30] "General Smuts Hopes Old Friends will Return to the Fold", *Rand Daily Mail*, 11 November 1940.

[31] "Cape Towns [sic] Welcome Steel Commando", *Rand Daily Mail*, 23 October 1940.

going to be useful after the war too". And the Steel Commando represented this new, modern warfare and a modern South Africa: "They know they've got modern, mechanical equipment at least equal to any European power's – and they're showing the world they know how to use it".[32]

As the Commando made its way back to Johannesburg, it stopped briefly at Doornkloof where its commanding officer, Colonel F. Cilliers placed a wreath on the grave of his namesake and Voortrekker pioneer, Sarel Cilliers.[33] Sarel was a key figure in Afrikaner nationalist history who had been one of those involved in the Battle of the Blood River against the Zulu in 1838—an event that had been commemorated with the recreation of the Great Trek a century later. As most South African schoolchildren would have learnt, he was also associated with the "vow", a promise made to God that should the Boers triumph over the Zulu, they would observe the day (the 16th of December) as a holy one. It was a founding story in Afrikaner nationalist mythology that symbolized the belief that the Boers—or Afrikaners—were God's chosen people. With the return of the Steel Commando just a week before the Day of the Vow (as it was then termed),[34] the men of the Commando were described as "triumphant Trekkers".[35] And Smuts would reinforce the association in his speech, while simultaneously referring to another key moment in Afrikaner nationalist history—the South African War: "It has been an arduous and exacting trek – of course, not as arduous as that trek which I and my mounted commando of young heroes made across South Africa about 40 years ago, but still very arduous and making great demands on all ranks".[36] In a climate where faced with rampant and resurgent Afrikaner nationalism that threatened to derail the war effort and Smuts's vision of unity, the South African Prime Minister and Commander-in-Chief of the UDF placed the Steel Commando firmly within the historical narrative of Afrikaner nationalism, a narrative that he now attempted to appropriate and employ for the War.

32 "We're not reading history…we're making it!", *Rand Daily Mail*, 10 December 1940.

33 "Wreath Laid on Grave of Trek Leader", *Rand Daily Mail*, 10 December 1940.

34 In South Africa, the public holiday is now known as the Day of Reconciliation.

35 "Greet the Commando To-day", *Rand Daily Mail*, 10 December 1940.

36 "City Praised for Stirring Welcome", *Rand Daily Mail*, 11 December 1940.

In his address, Smuts also spoke of the reasons that lay behind the tour of the Steel Commando. The first was that, although there was a public awareness of South African warfare, this bore little relation to the tremendous "changes" that occurred in warfare over the past two decades and the public would need to be disabused of the notion of the UDF using "bushcarts". The Steel Commando was to expose citizens to "a modern, mechanized army unit". It was also to reassure citizens that the UDF was not sending its soldiers "unarmed, unequipped [and] ready prey" for the modern Axis armies, but that South Africa was more than capable of holding its own against European forces in East Africa. There was too, the assurance that the UDF—even if composed of volunteers—was a disciplined fighting force, "allaying rumours of it being a mere mob, undisciplined and rude".[37] The Steel Commando then was an early exemplar of a modern South African army, a disciplined and dedicated citizenry and the industrialization of a modern South African state.

The War Train

The press releases prepared by the UDF for the prospective "successor to the Steel Commando", the War Train tour, emphasized South African self-sufficiency in the War. The Train would carry the armaments, food supplies and other goods that "South Africa has made with her own hands and her own resources". At the envisaged thirty or so major stops—each of which would last for about a day—visitors were to be exposed to the "real size and scope of South Africa's Home Front effort"[38] (Fig. 3.1).

Fittingly, the start of the "War Train" tour began in March 1941 with "a big opening show" held at Milner Park, the original site of the Rand Show. The War Train was under the command of the Director of Recruiting, Colonel C.G.C. Werdmuller. It was scheduled to leave Johannesburg on the 24th of March after the grand opening show and then wind its way along the country's railway network, moving through Natal, the Orange Free State, the Cape and finally returning to the Transvaal. It was also, however, a demonstration of South African prowess in Africa and the train was scheduled to stop in Southern and Northern Rhodesia

[37] "Premier Praises City for Stirring Welcome to Steel Commando", *Rand Daily Mail*, 11 December 1940.

[38] The Bureau of Information, "The War Train", Union War Histories, Box 277, File B.I. 29: "War Train National Tour", DOD.

Fig. 3.1 Part of a Pamphlet depicting the Prospective Route of the War Train across the Union, Union War Histories, Box 277, File B.I. 29: "War Train National Tour", DOD

(Zimbabwe and Zambia) as well as the Belgian Congo, with an estimated journey length of almost fifteen thousand kilometres over twelve days. On its various stops there would be a "spectacular military display" with servicemen and women demonstrating the weaponry and military capability of the UDF as well as the screening of films showcasing South Africa's war effort. The key aim of the War Train tour was to foster support for the war effort and thereby increase enlistment, and this was done through the showcasing of South African industrial might. The first part of the train consisted of the products envisioned by the country's "best brains" who had been tasked with accelerating the pace of industrial production to meet the demands of war. This section of the train held machinery, weapons and ammunition including an eighteen-pounder gun, an armoured car complete with crew to show off its features to the crowds and the very "first howitzer" to be wholly built in South Africa (and which had been on display during the Rand Show). An army relied on more than just weapons and there were displays of the food produced by South African farmers, "canned under the most hygienic and modern conditions" to feed the troops on the front lines. The spectacle was designed to amaze, to instil a feeling of patriotism in South Africa's transition to modernization and to demonstrate the importance of home front production in the War.[39]

The preparation that went into this recruitment drive was as impressive as the industrial might showcased. The two sections of the train each held sixteen military vehicles, including armoured cars and trucks and on board were members of the Special Service Battalion, the Railways and Harbours brigade, the Operating Group and two military bands. The combined weight was such that, at various points on the line, the two sections of the train had to be divided into three. And the arrival of the train at the various stations was only the final part of a logistical exercise that necessitated "an advance party" to prepare each point of arrival: the setting up of staging areas for the displays, the provision of electricity, lighting and crowd control. Railway cars had also been adapted to serve as display sites through which spectators could walk and view the various exhibits, including "life-size models" of servicemen and women in the

[39] "The War Train: A Comprehensive Tour of the Provinces", *South African Railways and Harbours Magazine*, Vol 35, April 1941, pp387–389 and "A Great Railway Achievement", *South African Railways and Harbours Magazine*, Vol 35, August 1941, p939.

various branches of the UDF. For those living on board the train for two weeks, there was the provision of water with supplies replenished frequently as well as food with an estimated one hundred thousand meals prepared over the course of the tour.[40]

In addition to the display of military material, the Train also showcased the transition made by the enlisted from civilian to military life. Aboard were six men, five of whom were from the Transvaal and none of whom had any prior technical expertise. They were respectively three farmers, a postman, a carpenter and an apprentice working on a grain elevator. All six men had enlisted in the Central Organization for Technical Training and, during the lengthy journey of the War Train, they were expected to continue their training under an instructor for the edification of the crowds. Detailed descriptions were provided of these men, along with the opportunities that service in the UDF would afford them in their post-war lives. 28-year-old J.C. Erasmus, for instance, had worked on his father's farm in Klerksdorp before the War. After his enlistment, he was trained in the skills of "fitting, blacksmithing and turning" that would be used during the War, and which would prepare him for either a career as a skilled worker or enable him to apply his new skills when he returned to the farm. S.F. Spies was a 17-year-old postman who found his job "dull". His enlistment came after he obtained parental permission and he was expected to have a far more stimulating future at the end of the War, due to the skills he had now acquired. The oldest of the six men was 37-year-old stock farmer, C.F.H. van Wyk. His technical training was to provide his wife and three children with "future security" at the end of the War. These detailed descriptions of the individuals travelling aboard the War Train were disseminated by the Bureau of Information and were to be provided to the various newspapers to be printed upon the arrival of the War Train in each region.[41]

If the War Train and its depiction of the modernization, militarization and industrialization of wartime South Africa was to serve as a site of aspiration and inspiration for potential recruits, it also presented a stark distinction from indigenous South Africans portrayed as traditional or pre-modern. At a stop in the town of Piet Retief, the train

[40] "A Great Railway Achievement", pp939–943 and "The War Train", p389.

[41] Bureau of Information, "War Train", Union War Histories, File B.I. 29: "War Train National Tour", Box 277, DOD.

was surrounded by “thousands of Swazi warriors in full war dress, armed with knobkerries and oxhide shields” who “performed a war dance”. The disjuncture between the modern and traditional was heightened with the sight described as “a striking contrast between primitive and modern weapons”. A message from Jan Smuts was subsequently relayed by the commanding officer of the War Train to the Swazi.[42] The Train thus served to incorporate the Swazi into the war effort, while nevertheless relegating them to a support or secondary role due to their incompatibility with modern warfare.

Midway through the Train’s tour of the Union—a journey described as the “longest train journey ever undertaken in South Africa”—the War Train was lauded in the English press as an unqualified success. This was particular evident by the reaction of those most likely to be opposed to the war effort, the denizens of the platteland or rural South Africa. As the Train stopped off in rural towns from Dundee in northern Natal to Bethlehem in the Orange Free State, spectators were given demonstrations of the various weapons, from the large 18-pounder gun to machine guns. They saw the enormous numbers of ammunition and helmets, parachutes and bombs. Farmers were gratified to see the woollen blankets produced from the wool of their sheep—showcasing their own role in the war effort. There was too a graphic description of the production of army boots, and people were given sufficient canned army rations to feed a family along with “miniature army biscuits” produced solely for the War Train.[43] The products of modern war and mass industrialization were brought to the country’s heartland and, as much as they were used as an attempt to foster nationalism and unity, “the bewildering array” of items on display also served as a reminder of the power of the warfare state.[44]

Just as the Train served as an illustration of a modern war effort for South Africans, it would do the same for the country’s neighbours. Upon the return of the War Train to a rapturous welcome in the capital city, Pretoria, Werdmuller described the reception received by the Train

[42] Dominions Office Press Section, “War Dance round ‘War Train’”, 5 April 1941, Union War Histories, Box 277, File B.I. 29: “War Train National Tour”, DOD.

[43] “War Train Proves Popular”, *Cape Times*, 12 April 1941, Union War Histories, Box 277, File B.I. 29: “War Train National Tour”, DOD.

[44] “War Train in City on Friday”, *Cape Argus*, 16 April 1941, Union War Histories, Box 277, File B.I. 29: “War Train National Tour”, DOD.

in southern and central Africa. For the Rhodesians, there was astonishment that South Africa's war effort had been the work of volunteers.[45] Volunteerism however was not simply a product of patriotism but had been a strategic decision on the part of the state, given the divided support for the War. Upon the Train's arrival in Umtali (present-day Mutare in Zimbabwe), Werdumuller read a message from the South African Prime Minister where Smuts extolled the part the Rhodesians had played in the War "alongside the South Africans". Yet the Train was also a means of showcasing "some of the great developments" made in South Africa "under the present stress of necessity".[46] This sense of South African prowess was further emphasized upon the Train's arrival in the Belgian Congo. Speaking in Elizabethville, the Governor of Haute Katanga spoke of the Train's role in heightening the "esteem" held by the Belgian Congo towards their southern "neighbours".[47] And, although not a part of the War Train, Smuts's wartime popularity was omnipresent, with him being referred to "as the Marshal of Africa".[48] Not just the South African leader, he was portrayed as representing the continent, just as the War Train represented the Union's military and industrial capability—and dominance. Werdmuller's reply to the foreign receptions was a gracious one—that also alluded to South Africa's future role on the continent: "Meeting neighbours and making closer ties of friendship can only result in closer cooperation in the future".[49] Werdmuller's words can be contextualized by the notion of a "Greater South Africa". As Hyam and Henshaw discuss, from the late nineteenth century, Smuts had articulated South African imperial aspirations in Africa with a re-envisioning of the country's borders from the Cape to the equator and including Rhodesia and parts of Mozambique. While these related to Smuts's—and succeeding leaders'—territorial ambitions, there was also an even more extensive notion of a South African sphere of influence that

[45] "The War Train Returns", *The Star*, 6 June 1941, Union War Histories, Box 277, File B.I. 29: "War Train National Tour", DOD.

[46] "S.A. War Train Arrives at Umtali", *Rand Daily Mail*, 17 May 1941, Union War Histories, Box 277, File B.I. 29: "War Train National Tour", DOD.

[47] "Messages to War Train", *The Star*, 5 June 1941, Union War Histories, Box 277, File B.I. 29: "War Train National Tour", DOD.

[48] "The War Train Returns".

[49] "Messages to War Train".

would comprise "economic and political hegemony".[50] The Director of Recruiting was therefore tapping into an existing perception of South African dominance that had been illustrated by the War Train.

For South Africans however, the arrival of the War Train had a mixed reception. The Mayor of Stellenbosch, Louis Hofmeyr felt inspired enough to enlist as a private in the Tank Corps on the day the Train came to town.[51] Yet the village council in the small town of Nylstroom, just a little more than a hundred kilometres from the Union's capital, Pretoria, refused to acknowledge the War Train and take part in any activities associated with its arrival.[52] As such, the tour of the War Train—as with the other attempts to recruit and to foster support for the war effort—also revealed the fractures in South African society.

The Air Commando

A few months after the War Train returned home, Werdmuller would take to the skies with the Air Commando. As Director of Recruiting, he had an unenviable job during the Second World War. Born in 1891, George Werdmuller served during the First World War and resumed his military service as a Captain in 1934, serving for the next five years in the Regiment, Westelike Provinsie. Just four days after the outbreak of the War, he was appointed Director of Recruiting and in June 1942, was also made Director of the Women's Auxiliary Army Services (WAAS).[53] His role in recruiting made him indispensable. In the aftermath of the War Train tour, he was hospitalized for a "gastric haemorrhage". Upon release, he relinquished his allocated sick leave, due to the "important work" in which he was engaging.[54] This would be the Air Commando.

From as early as 1923, Pierre van Ryneveld lauded the virtues of the air display. It brought the pilot into proximity with the civilian who would

[50] Ronald Hyam and Peter Henshaw, *The Lion and the Springbok: Britain and South Africa since the Boer War* (Cambridge, Cambridge University Press, 2003), p103.

[51] "Stellenbosch Mayor Joins Up as Private", *Rand Daily Mail*, 25 April 1941.

[52] "Nylstroom Council to Shun War Train", *Rand Daily Mail*, 3 April 1941.

[53] "Record of Service of George Clement Gie Werdmuller", File No R/6700: Werdmuller G.C., 87352V, DOD.

[54] Letter by H.S. Roseman, SAMC, to the Adjutant General, "N/N: Colonel G.C.G. Werdmuller, D of R", 20 October 1941, File No R/6700: Werdmuller G.C., 87352V, DOD.

potentially become a comrade. For the civilian spectator, there would develop an appreciation for the challenges of flying. Most significant, it served as a training exercise, providing valuable experience for pilots engaged in cross-country flying, developing their directional and aircraft maintenance skills.[55] By 1941, the South African Air Force (SAAF) had just turned twenty-one and the Air Commando was to commemorate the event. The launch was at Germiston, a star-studded event that included Van Ryneveld who had been tasked with forming the SAAF two decades earlier, as well as Jan Smuts, who had taken a particular interest in air power and spoke proudly of the expansion of the SAAF from 1,500 men at the outbreak of the War to its current capacity of 25,000 men (and women in its auxiliary division, the Women's Auxiliary Air Force). Speaking before a crowd of 25,000 civilians, Smuts praised South African efforts in playing "catch up", believing the SAAF would play a key part in achieving victory. The Air Commando was to "carry a message of goodwill and encouragement, of faith and hope, to every corner of South Africa". Its role was to include the people of South Africa as part of a "great army of liberation, liberty and freedom" against the totalitarianism of Adolf Hitler.[56]

Smuts's stirring words were followed by extravagant aerial displays by both aircraft and a glider. To demonstrate the accessibility of flight, a flight lesson was part of the display, with the instructions given to the student broadcast to the spectators. On the ground, the WAAF carried out a "smart drill" followed by a mock battle with armoured cars. There was a re-enactment of an "air raid" for an audience distant from the front lines and the air show culminated in a "fly past" of the various aircraft that comprised the SAAF to showcase "the rapidly expanding air arm".[57] With this spectacular beginning, the Air Commando took flight, travelling the length and breadth of the Union.

This was not the first campaign carried out to recruit pilots for a global conflict. As Tilman Dedering shows, over the course of the First World

[55] Pierre van Ryneveld to Chief of General Staff, UDF, "Aircraft Co-operation with Infantry: Bivouac 7th Infantry", 10 March 1923, Chief of General Staff, File C.G.S. 391/56/1, Vol 4: "Co-operation with Ground Troops and Navy; Co-operation with A.C.F. Units", Group 2, Box 260, DOD.

[56] "Thrilling Aerial Entertainment: Air Commando's Pageantry", *South African Railways and Harbours Magazine*, Vol 35, September 1941, pp1109–1110.

[57] "Thrilling Aerial Entertainment", p1111.

War, a South African pilot, Major Allister Macintosh Miller with Britain's Royal Flying Corps (RFC), was tasked with recruiting four hundred men for the RFC. Yet, even as pilot training was in its infancy, there was a clear delineation of who could constitute this elite club with Macintosh being ordered to recruit only those "of pure European descent". Yet, even among the "pure", there was some ambivalence, if not outright opposition, to South Africa's involvement in the War a little more than a decade after the end of the South African War. In Mackintosh's recruitment tour of 1918, of the more than three thousand applications received, twenty-five were from "Nationalist boys". Smuts's vision of the use of air power was one that was very much tied to domination—Miller's "aerial displays" were to quell potential "restlessness" among both Africans and Afrikaners. Air power would be more explicitly demonstrated during the uprising by white miners in the Rand Strike of 1922, when Smuts ordered bombing raids on Johannesburg, ultimately quelling the strike with casualties that ran into the hundreds. That same year, South African air power was deployed against the Bondelswart Nama in South West Africa, a South African mandate. Yet, just as air power was used to subdue, it was also a means of inclusion with "recalcitrant…Afrikaners [exchanging] their ox wagons for aeroplanes".[58] The use of "commando" in "Air Commando" then during the recruiting campaign of the Second World War, just as had been the case with the "Steel Commando", was to suggest the incorporation of the modern within the traditional, to appeal to a sense of history and Afrikaner nationhood while simultaneously serving to caution and intimidate the far-right elements that threatened the war effort—but also conversely created the public space where South African strength and unity could be challenged.

The prospective arrival of the Air Commando was heralded by a publicity campaign, including advertisements taken out in newspapers, one of which proclaimed: "The Air Commando is coming" and giving citizens the opportunity "to see South Africa's victorious Air Force in action…an Air Force to be proud of". The jubilant tone was accompanied by an illustration of planes zooming above the clouds as they converged upon a "For Victory/Vir Vryheid" crest of the SAAF. There

[58] Tilman Dedering, "Air Power in South Africa, 1914–1939", *Journal of Southern African Studies*, Vol 41, No 3, 451–465, 2015, pp453–457.

was too a map of the Union, detailing the route of the Commando.[59] Between August and mid-October—when it would return to Pretoria—the Air Commando was expected to stop at a number of towns across the Union: places such as Ermelo, Witbank and Standerton in the Transvaal; Newcastle in northern Natal and the urban centres of Pietermaritzburg and Durban; Bethlehem and Bloemfontein in the Orange Free State; and stopping throughout the northern, eastern and western Cape. Each stop would last a day to maximize the opportunity for the public to observe the flying acrobatics and the accompanying military displays.[60] In the ambitious display held at the Germiston aerodrome, for instance, at the outset of the tour, a dramatic re-enactment of an air raid was enacted. Buildings were constructed just for the event with a four-storey building set on fire by incendiary bombing. As members of the Civilian Protection Services attempted to douse the blaze, they were subjected to mock attacks by an Italian Caproni bomber which was subsequently engaged in an "aerial dog fight" by British-made Hawker Furies—all for the entertaining edification of "thousands of spectators".[61] Other displays were less ambitious but technologically impressive: Italian bombs used during the Abyssinian campaign (which was still ongoing), a model of a cockpit depicting how bombs were launched and a working field radio. A Gypsy Moth biplane was on display to illustrate the "structure of a plane" along with two engines—one an American Pratt and Whitney Wasp and the other the famed Rolls Royce Merlin used in the iconic British Spitfire. The "air photographic section" demonstrated the techniques of aerial photography and the developing of film. The complex technology on display was associated with youth—both the relative youth of the SAAF which had just turned twenty-one as well as the country's youth, who were most energized by the newest branch of the UDF.[62]

While emphasizing youth, the Air Commando also promoted the sense of unity so essential to the war effort with displays and exhibitions by the Women's Auxiliary Air Force (WAAF) as well as workers installing engines.[63] Speaking at Vryburg, Werdmuller encouraged women to enlist

59 "The Air Commando is Coming", *Rand Daily Mail*, 2 August 1941.

60 "Air Commando to Stage New Flag Ceremony", *Rand Daily Mail*, 5 August 1941.

61 "Real Bombs to be Used in C.P.S. Display", *Rand Daily Mail*, 6 August 1941.

62 "City's Big Welcome to Air Commando", *Rand Daily Mail*, 7 August 1941.

63 "City's Big Welcome to Air Commando".

in the WAAS or the WAAF. He spoke of their role in a "serious" conflict where it was necessary for them to relieve men for service on the front-lines. While downplaying the possibility of women being placed in danger, the Director of Recruiting nevertheless drew upon the historical imagery of "Voortrekker women [fighting] side by side with their menfolk".[64] Evident too was the incorporation of South Africa's neighbours. The Mswazi aerodrome in the independent kingdom of Swaziland (present-day Eswatini) was inaugurated on 15 August 1941 when the Caproni bomber which was part of the Air Commando became the first to land there. The construction of the aerodrome was both a symbol of Swazi "loyalty" to the Union (and the Allies) and a product of the traditional and the modern where oxen were used, and Swazi men volunteered their labour to construct the aerodrome.[65]

Perhaps most important in a divided country, was the need to appeal to the residents of the rural towns of the platteland. Arguably the Air Commando's greatest success came in the town of Upington in the northern Cape. In 2011, the town bordering the edge of the Kalahari Desert, had a population just under 75,000. Seventy years earlier, some twenty thousand people gathered to see the Air Commando. Described as the "biggest crowd which has ever gathered in Upington for any single event", the awe-inspiring display so moved a resident that he forewent his membership in the fascist Ossewabrandwag (OB), claiming a newfound allegiance to the "Government". In another town, another OB member enlisted in the UDF.[66] The Air Commando then—through its use of the spectacle of technology and the awe evoked by aircraft—was designed to both uphold state power and foster a sense of national unity. In stark contrast, a letter distributed to the various newspapers in the country and accredited to a leading figure in the OB, the Reverend C.R. Kotze, warned that: "The unity of which they boast in South Africa is a false unity, for which we are all paying dearly. It will not endure for long".[67] Werdmuller, too, had to respond to censure regarding the cost of events such as the Steel Commando, the War Train and the Air Commando with the declaration that their role in "recruiting and war propaganda"

[64] "Women Recruits Needed", *Rand Daily Mail*, 8 October 1941.

[65] "Caproni First Plane on New Aerodrome", *Rand Daily Mail*, 16 August 1941.

[66] "Air Display in Pretoria on Saturday", *Rand Daily Mail*, 15 October 1941.

[67] "Ex-Ossewa Chairman's Letter", *Rand Daily Mail*, 10 September 1941.

was invaluable as was the "deepest impression" they made on rural South Africans. For Werdmuller, "They are doing more to draw together our own people than anything else that I can imagine".[68] The Air Commando would therefore not be the last, nor even the most ambitious, attempt by the UDF to use technology to foster support and "unity" for the war effort.

The Liberty Cavalcade

A number of Liberty Cavalcades were held all over the country for the duration of the War with exhibitions taking place in Johannesburg, Cape Town and Durban. It was a month before the ignominious fall of Tobruk that saw thousands of South Africans taken prisoner, yet the Allies were already on the back foot in May 1942. While the United States had joined the Allies six months earlier, the Japanese were running rampant in the Pacific and both Singapore and the Philippines had fallen to Japanese forces; the Germans were continuing their advance into the Soviet Union while confronting the Allies over the island of Malta. It was within this context of Allied setbacks that the Liberty Cavalcade took place in Johannesburg in May 1942. Opening on the 23rd of May—a day before the Commander-in-Chief Jan Smuts's 72nd birthday—the Cavalcade was scheduled to run until the 1st of June.[69]

The Johannesburg Cavalcade was envisioned as a "wonder city" with an emphasis on unity in the face of adversity. Exhibitions were therefore wide-ranging and included those by other Allied countries, the mining houses, the South African Women's Auxiliary Services and those involved in the production of armaments and other equipment for the war effort. There was too an emphasis on South Africa's past, both industrial and military—and visitors could view the history of South African rail and transport, the process by which gold was extracted from ore and, of special significance to Jan Smuts, an exhibition titled the "Outspan": "a replica of a Boer commando's laager", comprised of waggons set among thorn bushes that served as a reminder of the early settler history of South

68 "Commando Not a Waste of Money", *Rand Daily Mail*, 23 September 1941.

69 "Gates of Liberty Cavalcade to Open this Morning", *Rand Daily Mail*, 23 May 1942, Union War Histories, Box 278, File B.I. 29: "Liberty Cavalcade", DOD.

Africa.[70] Included was an exhibition of "beautiful bombs" with munition workers displaying their skills in the War Supplies exhibit and drawing attention to the invaluable contribution made by citizens on the home front.[71] An armoured train, complete with ordnance, was also on display as was a downed German Messerschmitt, "General Smuts's baton" and pictorial depictions of action in North Africa.[72]

Of relevance here—and despite the acknowledgement of the various nationalities, ethnicities and cultures that comprised the war effort (as discussed below)—the emphasis was on the industrialization and technology that comprised modern warfare and which was a hallmark of the South African state. A newspaper article accentuated the loyalty and "generosity" of "the African people" who, along with white South Africans, had forwarded gifts to the Cavalcade to raise funds for the war effort. Among these was a walking stick inscribed with "To my Government – one stick". More sticks were received from the Eastern Transvaal, carved with "guns, tanks and men in uniform". These were described as evidence of the changes wrought in the "material…and social culture of the people" fostered by the War. Another gift received could only be described as quaint—if not anachronistic: "A little shield, spear and knobkerrie were handed over as 'our weapons of war, and we are giving them for the war, which is to bring us victory'".[73] In a depiction of a war effort so defined by science, technology and mass industrialization and despite overt attempts to present an inclusive and united front, modern warfare nevertheless also served as an area of exclusion—the historically defeated enemy of white settlers whose weapons were superfluous items of curiosity in the current conflict.

Yet those considered auxiliary to the war effort were also incorporated into the narrative of technology and industrialization in particular ways. At the exhibition of the industrial work done on the home front, white women skilfully demonstrated their new roles, using "oxy-acetylene lamps with the same ease…[as that] of a frying pan". Another was tasked with

[70] "Gates of Liberty Cavalcade to Open this Morning".

[71] "'Beautiful Bombs' to be Exhibited at Liberty Cavalcade", *Rand Daily Mail*, 19 May 1942, Union War Histories, Box 278, File B.I. 29: "Liberty Cavalcade", DOD.

[72] "Union's Strangest 'City' Takes Shape at Zoo Lake", 11 May 1942, Union War Histories, Box 278, File B.I. 29: "Liberty Cavalcade", DOD.

[73] "Great Final Spurt at Liberty Cavalcade", 29 May 1942, Union War Histories, Box 278, File B.I. 29: "Liberty Cavalcade", DOD.

attaching rivets to land mines. Able to do two hundred a day, she pointed out that men, who were able to use a "stronger flame" could exceed her target. A "house-wife" was assembling material that would be used on "practice bombs" while others manufactured screws for guns. Those women with a particular aptitude for science and mathematics—and who had been teachers in 1939—were now lieutenants in the SAAF overseeing their junior colleagues in the maintenance of military aircraft.[74] The incongruous image of women working in the manufacture of weaponry and war material was therefore offset by drawing parallels and emphasizing continuities with domestic and conventional gendered divisions of labour.

Another highlight for visitors was the experience of being in a goldmine. The discovery of gold sixty years earlier had transformed the South African economy from a largely agrarian one to one based on rapid industrialization and technological development. For decades, the mining of gold had been a mainstay of the economy, only to be supplanted in the war years by secondary manufacturing. It was mining that had made an indelible impact on the social and political structure of the country. Now, for those South Africans who were unfamiliar with the mines, the Simmer and Jack mining company staged an exhibit to acquaint visitors with the experience of gold mining. It had taken "four Europeans and 14 natives" two months to construct and exhibit the safely replicated (and packaged) experience of underground mining. Groups of eighty were taken on an underground railway into the "bowels of the earth", there to encounter the "jackhammers drilling, water-pumping [and] miners at work". The exhibition was a huge drawcard with visitors waiting in long queues and a total of ten thousand viewing the exhibit in just two days.[75]

In addition to the displays related to the War and to industry, there was also a sense of whimsy. The site of the Cavalcade was Zoo Lake, and the exhibition covered more than a hundred acres with an amusement park, complete with "haunted tunnel and a crazy house" and a "Wild West rodeo".[76] In contrast to the rowdy frontier of the Wild West was

[74] "City's Remarkable Response to Liberty Cavalcade", *The Star*, 27 May 1942, Union War Histories, Box 278, File B.I. 29: "Liberty Cavalcade", DOD.

[75] "Cavalcade an Unqualified Success", *The Star*, 26 May 1942, Union War Histories, Box 278, File B.I. 29: "Liberty Cavalcade", DOD.

[76] "Cavalcade City Visitors will Need Maps", 15 May 1942, Union War Histories, Box 278, File B.I. 29: "Liberty Cavalcade", DOD.

a quaint English village, complete with inn. Children were entertained by the "Treasure Island"-themed rides and the possibility of finding a buried pirate hoard. A tree house was built to represent Wendy's home in "Never-Never Land" and youngsters could meet Peter Pan or walk the plank of the "Jolly Roger" and both children and their parents could ride on a miniature railway.[77]

The most significant displays however were those that emphasized the united nature of the war effort. In addition to the representation of the Wild West (the newest ally being the United States) and the English village, visitors were exposed to a veritable international tour with exhibits by Scotland, Czechoslovakia, Greece, India, Belgium, Denmark, Palestine and South Africa's sister Dominions, Australia and New Zealand. There was a "Maori village" and the replica of a "skin tent" from Lapland, complete with "Lap fortune-teller", as well as "a yogi's mystic haunt". The wide-eyed were also exposed to the cultures of the many nations that comprised the Allies—or who had already experienced the brunt of the conflict. This included Russian food, music and ballet, "Belgian Congo coffee" and an American-themed restaurant complete with cabaret.[78] Allied solidarity would reach its zenith with "Dances of the Allied Nations" to be held on the 29th of May. Comprising 250 participants representing fifteen nations, dancers would be attired in "national costume" and accompanied by the UDF band.[79] On opening the Cavalcade, Smuts was presented with a "golden dagger" by the representatives of the Indian exhibition—an exhibition that included the products of subcontinental craftsmanship from Tibet, Nepal and India. Closer to home, "native work" derived from Northern Rhodesia (Zambia) was available for purchase. The "native section" set up by the Municipal Native Affairs Department consisted of a reconstruction of a village with afternoon exhibitions of children singing and dancing followed by "war dances" by "warriors" in the evenings. The head of the exhibition, a

[77] "Union's Strangest 'City' Takes Shape at Zoo Lake".

[78] "Union's Strangest 'City' Takes Shape at Zoo Lake" and "Panorama of Nations at Zoo Lake City", 16 May 1942, Union War Histories, Box 278, File B.I. 29: "Liberty Cavalcade", DOD.

[79] "Great Final Spurt at Liberty Cavalcade".

Mrs Graham Ballenden, was "[gratified at] European interest in our section".[80]

Yet, despite this idealized representation of international and domestic harmony, the tensions of the War—and those manifest in South African society—were evident when just three days into the Cavalcade, the Danish stand was destroyed by an incendiary bomb that caused a fire. Additional concealed explosives were found at the Soviet pavilion and throughout the exhibition grounds.[81] The bomb had been discovered by a "native police boy" and was one of a number that were timed to go off, creating fires with the aim of destroying the "wonder city". Police investigators concluded that it was a "well-planned attempt by a number of saboteurs".[82] The sabotage notwithstanding, crowds continued to flock to the Cavalcade with another visit by Jan and Issie Smuts in a show of solidarity. The total attendance at the Cavalcade over its ten-day run was 377,273 with 28,000 attending the day after the attack.[83] Yet, the indubitable success of the Liberty Cavalcade in 1942, was overshadowed by the Cavalcade held in Cape Town two years later—which adopted a confident and far more martial tone.

Despite the setbacks of 1942, by 1943, the tide of war had clearly turned in the Allies' favour with the defeat of German forces in North Africa, an Allied advance into Italy and the Soviet triumph on the Eastern Front. By 1944 then, just two months before the landings at Normandy that would ultimately spell an Allied victory in Europe, the Liberty Cavalcade was held in Cape Town in a very different climate. The timing of the Cavalcade was not happenstance—it took place while Parliament was in session[84] with Members of Parliament who visited the exhibition tasked with informing their "constituents" of the importance of a war that had

[80] "Indians to Give Rare Golden Dagger to Smuts", *Rand Daily Mail*, 22 May 1942, and "City's Remarkable Response to Liberty Cavalcade", Union War Histories, Box 278, File B.I. 29: "Liberty Cavalcade", DOD.

[81] "Incendiary Bomb Destroys a Cavalcade Pavilion", *The Star*, 26 May 1942, Union War Histories, Box 278, File B.I. 29: "Liberty Cavalcade", DOD.

[82] "Incendiary Bombs at Cavalcade", 27 May 1942, Union War Histories, Box 278, File B.I. 29: "Liberty Cavalcade", DOD.

[83] "City's Remarkable Response to Liberty Cavalcade" and "Cavalcade Closes", *Rand Daily Mail*, 2 June 1942, Union War Histories, Box 278, File B.I. 29: "Liberty Cavalcade", DOD.

[84] Cape Town is the legislative capital of South Africa while Pretoria is the administrative capital. The city of Bloemfontein is the judicial capital.

been ongoing for more than four years. The exhibition was designed to bring the reality of war home to those on the home front who had only been exposed to it via newspapers, newsreels and radio broadcasts and who would now be made aware of its "full significance".[85] To comprehend its import was to understand what modern war entailed and to sustain public support in a war fought for liberty—which Smuts believed would have historical meaning for South Africans:

> We entered this war in a solemn spirit of devotion. We were moved by the thoughts of a large menace. To us a new wave of barbarism seemed to be sweeping over the world. This young nation, born in liberty, accepted the call. It is still playing its part and, after four and a half years, means to play its part to the very end.[86]

The ubiquitous G.C.G. Werdmuller was the chair of the UDF organizing committee for the Liberty Cavalcade which was to be held from the 25th of March to the 1st of April 1944, at Green Point Common in Cape Town. Envisaged as the "most spectacular and outstanding military review" held in South Africa, the proceeds of the Cavalcade were to be donated to the Governor-General's War Fund. Jan Smuts had decreed the holding of the Cavalcade and subsequently opened the event, accompanied by a twenty-one-gun salute. The exhibition was intended to showcase all aspects and activities of the various armed services that comprised the UDF, including those that had hitherto been insufficiently publicized. These included the work of the Pay Corps responsible for payment of salaries as well as the Army Post Office. The Non-European Army Services were also included with "demonstrations of trick motorcycle riding". The overall effect was to give the public insight into "the workings of a modern scientific army"[87] (Fig. 3.2).

Instead of static exhibitions, visitors experienced "mock" attacks and defence, demonstrating UDF arms and armaments and the capability of its men and women. The repelling of a sea invasion of Cape Town via

[85] Lieut (Mrs) P. Douglass, "Cape Town Cavalcade in Retrospect", *The Nongqai*, May 1944, p549, DOD.

[86] "Great Crowd at Cape Cavalcade", *The Star*, 3 April 1944, Union War Histories, Box 278, File B.I. 29: "Liberty Cavalcade", DOD.

[87] Lieut (Mrs) P. Douglass, "Progress of Modern Warfare: The Army in Review", *The Nongqai*, March 1944, pp283–284, DOD.

Fig. 3.2 Cartoon entitled "Cavalcade Complications" depicting the logistical challenges of the Liberty Cavalcade, *The Nongqai*, March 1944, p301, DOD

Table Bay was staged by the Coastal Artillery. Anti-aircraft and machine guns bombarded the Atlantic, seeking out targets illuminated by searchlights. For dramatic effect, the demonstration was held at night, with spectators treated to the "vivid flash of orange, scarlet and yellow which turned the whole sky to flame" as tracer bullets lit up the sky, accompanied by the "ear-shattering boom and crash of the guns". Both the Navy and the Air Force joined in, symbolizing the spirit of co-operation between the various branches of the UDF. For those unable to be there in person, detailed commentary was provided by the South African Broadcasting Corporation (SABC) to citizens listening all over the Union from 19:30 to 22:00. It also marked for many an introduction to the newly established UDF Radio Station, operating on both longwave and shortwave bandwidths to reach the entire Union. Regular reports from the Liberty Cavalcade under the Directorate of Military Intelligence were also printed and distributed as a UDF newspaper, the *Union Star*.[88] For Smuts, who gave the inaugural address of the UDF Radio Station, the transmission of information was less about propaganda than it was about the fostering of "democracy" through "education" and the creation of a "better" post-war world. UDF Radio then was to use the reach of media "to give clarity to our ideals", to promote a particular vision of post-war South Africa that had been forged by the Second World War.[89]

The city described as the "Mother City" held a particular resonance for South Africans—it was the site of the first white settlement in the mid-seventeenth century, and the possible threat of foreign invasion therefore took on added significance. In a war that was fought in far-off theatres, civilians could certainly have experienced a sense of complacency, of safety from a war that was experienced largely through newspapers and radio with the "tragic significance" of these reports muted through repetition and over time. The demonstration then, with its assault on the senses and the portrayal of South African military might, was a means of bringing the war home, of both reassuring South Africans of the protection provided by the UDF and also of presenting a realistic image of the experience

[88] Douglass, "Cape Town Cavalcade in Retrospect", pp549–551.

[89] "Smuts Opens UDF Radio Station", *Rand Daily Mail*, 20 March 1944, Union War Histories, Box 278, File B.I. 29: "Liberty Cavalcade", DOD.

of war. Onlookers were expected to be awed, to be "impressed by its grandeur and by its dramatic importance".[90]

Underpinning the demonstration was the military technology employed to such great effect by the UDF during the War and in addition to the theoretical defence of the home front, was a demonstration of UDF tactics on the battlegrounds of Europe. The Engineer Corps built "a model village" in the twenty-acre arena, complete with "mine-fields, tank traps and barbed wire entanglements" and protected by hostile aircraft. To render the scene as realistically as possible, a "river" was also created, complete with water supply. Serving effectively as an obstacle course, the various branches of the UDF were then expected to showcase their talents in taking control of the village. First into the fray was the reconnaissance team who were then succeeded by the Engineers working to clear the minefields while supported by the unceasing bombardment of artillery, infantry and tanks. The infantry was then tasked with finding means of fording the river and then engaged in close combat with village defenders. UDF aircraft also jousted with the "enemy" in the skies while "softening up the village defences" and eventually harrying the defeated enemy. The aftermath saw the arrival of the Medical Corps, tasked with treating battlefield casualties. The dramatic enactment was for more than just public entertainment—it also served as a training ground for recruits, exposing them to battlefield conditions before they were sent to Europe.[91]

For those familiar with the recent events of the War, there was a depiction of a bombing raid on Berlin, part of a series of air raids that had been carried out by British and American forces from November of the previous year (and would only end in March when the Liberty Cavalcade took place). Based on an aerial photograph, a model was constructed of the German capital at the behest of the Directorate-General of War Supplies. To add to the sense of realism, city landmarks were depicted in the model, which stood at just over one square metre: as well as the factories and warehouses, there was the Opera House and the Cathedral. To simulate the "searchlights, flak and crashing bombs" associated with an air raid, there was the creative use of "nothing more terrifying than three motors, electric globes, coloured transparent paper, fowl feathers and wisps of cotton wool". The dramatic depiction served as another

[90] Douglass, "Cape Town Cavalcade in Retrospect", p549.

[91] Douglass, "Progress of Modern Warfare", pp284–285.

means of reminding civilians of South Africa's contribution to the war effort, with the note that the country's factories could potentially supply sufficient bombs for twenty-five bombing raids on Berlin.[92] It was therefore South African industrial and technological strength that played an integral role in the fight against the Axis.

An essential feature of the Liberty Cavalcade was the display and history of war technology. Evident was an exhibition of British and American tanks which had been used by the UDF on the frontlines. Accompanying this was a demonstration of the Armoured Recovery Unit who worked on "damaged tanks", giving the public an inkling of their work up north. There was also a display of the history of the development of the tank beginning with the British Whippet used on the western front in 1917. While the tank displays leaned heavily towards British and American invention and workmanship,[93] on display too was an example of South African ingenuity, albeit of a more prosaic kind. A system of "mud winches and cables" used to liberate heavy armoured vehicles—such as tanks—from "bogged terrain" was demonstrated and described as "a South African invention". The various weaponry used by the infantry—and their development into the modern weapons of the Second World War—was also part of the exhibition as were the methods by which bridges were built to ford rivers. Less dramatic—but no less essential—was the filtration system used by UDF engineers to purify drinking water in East and North Africa.[94]

Food was given even more publicity than water. Holding true to the adage of an army marching on its stomach, the domesticity of food production was infused with the science of nutrition of the modern UDF. The traditional army cook worked at the behest of various experts in nutrition such as "dieticians" and "research chemists" to provide a balanced diet for serving men and women. The UDF Cookery School had been established to train cooks in the preparation of meals for troops on the frontlines and within the Union itself. For the Cavalcade exhibit, those attached to the Cookery School worked in conjunction with the

[92] "Berlin Air Raid Shown in Miniature", *The Star*, 17 March 1944, Union War Histories, Box 278, File B.I. 29: "Liberty Cavalcade", DOD.

[93] Chapter Four discusses the South African invention of the flail tank—and the controversy surrounding this invention—which would be used a few months after the Liberty Cavalcade during the invasion of Normandy.

[94] Douglass, "Progress of Modern Warfare", p287.

specialists of the South African Medical Corps (SAMC) to demonstrate the preparation of food. These kitchens ran constantly, providing food for the UDF members attached to the Cavalcade while simultaneously demonstrating the art of food preparation in a "scientific modern army" to the public. Yet, while focusing on the feeding of an army, the principles of scientific nutrition were also considered to be useful for a modern and uniquely South African civilian society as well. With experts at hand to dish out "advice" on "cooking problems", there was also the consideration of the consumption of "cheaper" but healthy food; the adjustment of eating portions in relation to the country's climate and, significantly, the ideal diet for "non-Europeans". This had been determined from the "invaluable experienced" gained in providing for the men of the Non-European Army Services (NEAS). In contrast to the conventional staple of "mealie [maize] meal", UDF nutritionists provided advice to "employers of labour" on how to provide their workers with a "varied" and cost-effective diet, designed to "enable them to work much more efficiently and happily".[95] The lessons of war then, would be applied to a reconstructed post-war world and be used to make a modern South Africa—albeit one based on existing hierarchies.

The exemplar of modern warfare was, of course, air power and the Air Force display at the Cavalcade followed in the tradition of the Air Commando held in 1941. Displays by the SAAF gave a dramatic visualization of the War to astonished onlookers, from simulated invasions to aerial combat. However the SAAF exhibition was far more extensive and ranged from demonstrations of the work of mechanics in maintaining the aircraft, the communication systems used, the testing of potential candidates and the medical science behind flying at altitude. Underpinning the exhibition of the SAAF was a sense of pride in the development and expansion of this most modern branch of the UDF with the acknowledgement of the "anachronistic" aircraft and the modest scale of the SAAF at the onset of the War where South African pilots had to be trained overseas due to lack of capacity. It was however through the leadership of "men with vision" and the support and commitment of pilots and groundcrew that the SAAF was transformed. South African pilots were now making a substantial impact on the war effort, holding their own

[95] Douglass, "Progress of Modern Warfare", pp288, 292.

against the German Luftwaffe. The SAAF was presented as the amalgamation of individual "courage"—a "South African tradition"—and the "skill and science" due to industrialization, training and technological development. The work of the SAAF during the War was described as "the modern counterpart of the Great Trek" as they fought the enemy in East and North Africa and Southern Europe.[96] Like the Air Commando then, the role of the SAAF in the War was placed within an existing historical tradition that nevertheless incorporated the modern—and, in fact, exemplified it.

From 1940 then, the War was brought to the South African public through various tours and exhibitions staged across the Union. There was, however, a realization that the War was an historic event, one that needed to be remembered and commemorated for posterity. And it was even as the dramatic events of the Liberty Cavalcade were unfolding in 1944 that mention was made of the War Museum which would preserve not just the historical events of the War itself, but its "tangible objects", the military technology that would serve as part of the historical record.[97]

The South African National War Museum was opened to the public in April 1947. Its first premises were temporary—at the Johannesburg Zoo and close to the War Memorial. While it would later incorporate other South African conflicts, its initial focus was the war that had just ended. Spread out over two halls were weapons (of both Allied and Axis origin), aircraft (including a Hartbees plane used in East Africa) as well as a model of a rescue boat used by the SAAF at the Cape. Other larger items, including a German plane, were displayed outside.[98] Another example of German technology would arrive a few months later when a German submarine was donated by the British Admiralty and was subsequently transported from Durban. A two-man submarine, it weighed more than ten tonnes and joined the military vehicles and equipment that dotted the grounds of the museum.[99]

As with the displays during the War, the museum was a site for the public to experience and explore military technology—and to see at first

[96] J. Sack, "The Background to Air Power", *The Nongqai*, March 1944, pp297–302, DOD. The work of Ernst Jokl in aviation medicine and high altitude flight is discussed in Chapter Seven.

[97] Douglass, "Progress of Modern Warfare", p287.

[98] "War Museum Opened in Zoo Grounds", *Rand Daily Mail*, 10 April 1947.

[99] "Museum Piece", *Rand Daily Mail*, 17 December 1947.

hand the weaponry that had defined the largest military conflict in history. After a war that had necessitated the UDF and the South African government calling for the unity of all South Africans, the museum nevertheless reflected the reality of segregation with different "opening hours" for "Europeans" and "non-Europeans" who were clearly to marvel at South African military prowess separately.[100] Complaints were also made about the unaesthetic nature of the hangars housing the exhibits, more so because "they [attracted] a large number of natives during weekends".[101]

Jan Smuts officially opened the museum in December 1947. In a speech brimming with patriotic pride at the admirable way in which South Africa rose to the occasion in September 1939, Smuts spoke of the paltry military equipment held by the UDF at the start of the War and how, within a relatively short space of time, South African forces were able to counter the Italians in East Africa. In addition to the "exploits of South Africa's fighting men", praise was given to those who had helped fuel the industrialization effort that had produced weapons and war equipment and who had provided supplies for ships stopping at the country's ports and harbours and for those in the Middle East campaign. Smuts accredited South African success to the hardy forebears of the contemporary generation, "the voortrekkers and pioneers". Yet, the museum was not the simple celebration of South African military might and industry as had been the case with the exhibitions during the War. After six years of warfare, sobered by the loss of millions of lives and devastation wrought on an epic scale, Smuts viewed the museum as a "memorial" but also as a cautionary tale where "we shall never have to use again the weapons of death and destruction that we see here to-day, or those still more dreadful to follow".[102] It was both a celebration of technology and a warning.

Five months later, Smuts was ousted from power in a national election that would bring the National Party to power and signal the victory of Afrikaner nationalists who—despite the demonstration of UDF military prowess during the War—had remained implacably opposed to South

[100] "War Museum Opened in Zoo Grounds".

[101] "City Needs Home for War Museum", *Rand Daily Mail*, 28 March 1947.

[102] "War Museum is Memorial to S.A.'s Greatest Efforts—Smuts", *Rand Daily Mail*, 30 August 1947.

African participation in the conflict. Under the new apartheid state, military technology was subordinated to a new ideology of domination, subjugation, division and exclusion.

CHAPTER 4

From Creation to Credit: Major Du Toit and the Invention of the Flail Tank

In 2016, a century after tanks first became the formidable weapons of the Western Front, a hitherto unknown poem was discovered that had been penned by First World War veteran and Winnie the Pooh creator, A.A. Milne. After experiencing trench fever during the War which ended his combat service, Milne produced propaganda extolling the British war effort. It is therefore in his role as a propagandist that his ode to tanks, simply titled "Poem", can be placed. Yet the poem also emphasizes human agency in this modern war with its emphasis on the overwhelming and brutal efficacy of technology:

> So remember, whenever you talk of the Tanks,
> The newest invention, the wonderful Tanks –
> The older invention – the men in the ranks...[1]

In contrast, for those confronted by tanks as powerfully depicted in Erich Maria Remarque's *On the Western Front*, they represented the very essence of the dehumanization of mechanized and modern warfare:

[1] Lucy McCalister, "A.A. Milne Poem", 9 September 2016 and A.A. Milne "Poem", *The Tank Museum*, https://tankmuseum.org/article/milne-tank-poem-found, accessed 21 December 2023.

S. Chetty, *Science and Technology in South Africa, 1939-1946*, Palgrave Studies in the History of Science and Technology,
https://doi.org/10.1007/978-3-032-05084-7_4

> We cannot see the gun batteries that are bombarding us, and the oncoming waves of enemy attackers are human beings just like we are - but tanks are machines, and their caterpillar tracks run on as endlessly as the war itself. They spell out annihilation when they roll without feeling into the shell holes and then climb out again, inexorably, a fleet of roaring, fire-spitting ironclads, invulnerable steel beasts that crush the dead and the wounded. Before these we shrivel down into our thin skins, in the face of their colossal force our arms are like straws and our hand-grenades are like matches.[2]

While the concept of the tank has a history as far back as Leonard da Vinci's sketches of an armoured vehicle in the late fifteenth century, it was the trench warfare of the Western Front that provided the impetus for its development.

The Somme is an area in northern France named after the river that flows through it. For much of the twentieth century, however, it had lost all geographical meaning, becoming instead synonymous with the conflict on the Western Front during the First World War. As Nicholas Murray discusses, the development of trench warfare had its origins in the late nineteenth century and was marked by the creation of extensive largely underground fortifications to protect men from the increasingly deadly effects of machine guns and artillery. Warfare took on a largely defensive aspect with the area—that would come to be known as No Man's Land—marked by barbed wire that hampered the advance of troops. The extensive trench fortifications of the First World War would make it all but impossible for the enemy to be flanked and hence over-run. The access to similar weaponry by both sides created the conditions for a stalemate.[3] Attempts at offensive warfare where the opposing fortification was first subjected to intensive artillery fire before soldiers advanced across No Man's Land resulted in a horrendous death toll, with little tactical advantage gained. This would be exemplified by the first day of the Battle of the Somme in 1916 which was marked by the deaths of almost twenty thousand British soldiers. Yet, as William Philpott argues convincingly, it was this lengthy battle—with the Battle itself lasting 141 days—that

[2] Erich Maria Remarque, *All Quiet on the Western Front*, Brian Murdoch (translator), (Alfred A. Knopf, 2018), p239.

[3] Refer to Nicholas Murray, *The Rocky Road to the Great War: The Evolution of Trench Warfare to 1914* (Washington DC, Potomac Books, 2013).

provoked necessary changes to the status quo. By 1916 in Britain, the industrialization of war had taken on a new impetus—and so too had the use of technology. Reconnaissance was carried out behind enemy lines using aeroplanes equipped with cameras; trench mortars and gas were used against enemy trenches; and the tank made its debut at the Battle of Flers–Courcelette that lasted a week, beginning on 15 September 1916. Undeterred by barbed wire or machine guns, the tank was a juggernaut designed to end "static" warfare.[4] Its brutal efficiency as a weapon of war meant that there was no shortage of claimants demanding credit for its invention.

The Royal Commission on Awards to Inventors was by no means unused to the controversy over inventions and patents. First established in the wake of the First World War in 1919, the Commission had been tasked with determining the inventors of the tank itself.[5] The tank would be an amalgamation of the existing technology dating back to the late nineteenth century: the internal combustion engine, guns mounted on turrets encased in armour and caterpillar tracks akin to those used on American tractors. It was no less prominent a personage than Winston Churchill, the First Lord of the Admiralty, who requested Commodore Sueter first to look at the feasibility of armoured cars in supporting aircraft mobilization by creating and holding airfields, and then to develop a method of permitting troops to traverse safely the barbed wire strewn mire of No Man's Land and reach the German trenches.[6]

Ernest Swinton joined the Royal Engineers in 1888 and served in the South African War. It was while in South Africa that he claimed to have been informed by a mining engineer about a "caterpillar tractor" that was used in the United States. Its ability to negotiate challenging terrain potentially made it a useful vehicle of war. Swinton would subsequently meet with his former commanding officer, Colonel Maurice Hankey in London in 1914. At the end of that year, Hankey drew up a memorandum describing an armoured vehicle, mounted on caterpillar tracks and equipped with a Maxim gun that would provide covering

[4] Refer to William Philpott, *Three Armies on the Somme: The First Battle of the Twentieth Century* (New York, Alfred A. Knopf, 2010).

[5] "Royal Commission on Awards to Inventors", *Nature*, No. 4013, 28 September 1946, p443.

[6] J.P. Harris, *Men, Ideas and Tanks: British Military Thought and Armoured Forces, 1903–1939*, (Manchester, Manchester University Press, 1995, 2015), pp9, 12–13.

fire and mow down the barbed wire that made No Man's Land such deadly terrain. Hankey's memorandum was, however, one of several ideas mooted for giving the British an advantage in trench warfare, ideas that included a "mobile armoured shield" that would be mounted on a single wheel, with caterpillar tracks eventually suggested to negotiate the muddy terrain. A far more ambitious idea was that of a "gigantic land battleship with 40-foot wheels", the development of which Churchill would allocate funding of £70,000. These "ships" would transport soldiers in relative safely across No Man's Land. By March of 1915, eighteen landships were commissioned, twelve of which would make use of early caterpillar tracks. Walter Wilson and William Tritton, both with engineering experience, were assigned to develop these landships.[7]

Slow progress was made—the envisaged size of the landship would make it cumbersome and heavy and thus susceptible to enemy fire or being bogged down in the mud. A field trip to France demonstrated that the narrow streets would prove extremely difficult for such a large vehicle to negotiate, and a recommendation was made for an "articulated" vehicle. By August 1915, there was a realization that caterpillar tracks presented a useful solution. Tritton and Wilson subsequently went on to produce the first prototype of an early tank, the Lincoln No. 1 Machine. Although it would require much adaptation and improvement, it marked the basic prototype of a tank: it was smaller; mounted on caterpillar tracks (which subsequently had to be refined); was "unitary" instead of articulated; and would not serve as a "troop carrier" but rather as an assault vehicle in its own right. The final product was the Mark I tank which would come in "male" and "female" variants. Swinton was responsible for the female version—a tank bristling with armaments that was designed to provide support for the heavier male version. These tanks made their appearance at the Somme on 15 September 1916.[8]

With the publicity accorded to these new weapons of war, came claims of invention, most of which were unfounded. Swinton, who had served as a war correspondent, was touted as the inventor by the press. Tritton and Wilson, however, had done the work of rendering the idea concrete and practical.[9] In December 1919, the Commission announced the awards for

[7] Harris, *Men, Ideas and Tanks*, pp13–19, 21–23.

[8] Harris, *Men, Ideas and Tanks*, pp24–26, 28–29, 31–33.

[9] Harris, *Men, Ideas and Tanks*, p33.

the invention of the tank. Out of a total of £18,000 awarded, £15,000 was shared between Tritton and Wilson with a far more modest £1,000 awarded to Swinton. Winston Churchill was singled out for his encouragement and willingness to embrace ingenuity, "his receptivity, courage and driving force", that realized the vision of the tank. Other figures involved in the development were acknowledged but did not receive monetary reward as they were considered to be "servants of the Crown" and therefore merely performing the duties to which they had been assigned. *The Spectator* took exception to the small grants which did not seem sufficient compensation for a seminal and war-changing invention; to the lack of award to the "servants of the Crown" as both public servants and civilians had played their part in the war effort; and, finally, to the lack of recognition accorded to the actual inventor of the tank, the Australian Lance de Mole.[10]

Yet the Australians could—and did—claim credit for the development of the tank before the British. In September 1914, G.W.D Breadon of Perth, Australia, wrote a letter to the War Office in London detailing the invention two years earlier of a "travelling caterpillar fort" by a Mr Launcelot de Mole. These armoured "forts" were envisioned as enhancing the capability of the infantry as they could traverse challenging topography and, with armour and the installation of weaponry, could "quickly turn the most stubborn of armies, even if they be strongly entrenched".[11] It was an eerily accurate image of the trench warfare that characterized the Western Front. De Mole was born in Adelaide, the son of a civil engineer. In 1912, the then corporal was working in "heavy transport" which likely inspired his creation. He proposed a design for an early prototype of the tank to the War Office but this was turned down. According to an Australian report, it was suggested that he send the proposal to the Germans instead, but he refused. He again submitted his proposal to the War Office in 1916, but it was apparently "pigeon-holed and forgotten". The Royal Commission on Awards to Inventors would subsequently conclude that De Mole had failed to demonstrate

[10] "The Awards Made to Tank Designers by the Royal Commission", *The Spectator*, 6 December 1919, http://archive.spectator.co.uk/article/6th-december-1919/3/the-awards-made-to-tank-designers-by-the-royal-com, accessed 18 December 2023.

[11] Letter by G.W.D. Breadon, "Travelling Caterpillar Forts", 19 September 1914, PRG 1319/1/1, State Library South Australia Digital Collections, https://digital.collections.slsa.sa.gov.au/nodes/view/2460, accessed 18 December 2023.

that the British had adopted his invention when creating the tank and that his great "misfortune" lay in inventing a device when there was as yet no need for it. A British newspaper put it best: "It is no use discovering the quinine until you have invented the fever".[12] While this is complicated by a sense of national pride in "being the first", the Commission clearly viewed invention in terms of its utility and relevance. As Tom Schactman suggests, the use of science and technology in war is often a reactive process, a response to real-world conditions.[13] Ultimately, De Mole's "failure" was about timing: there was no need for a tank in 1912, but trench warfare would make the invention invaluable less than five years later.

With the end of the First World War, it appeared as if the utility of the tank had come to an end, as evident in the words of Louis Jackson of the Ministry of Munitions, "The tank proper was a freak, the circumstances which called it into existence were exceptional and not likely to recur". Yet, the British Tank Corps continued in the interwar period and while circumstances may have been unlikely to repeat themselves, the tank could itself be adapted to new conditions of war. While increasingly subordinate to the growing dominance of air power, tanks in the form of armoured cars were deployed in the various hotspots of the British Empire.[14]

As Harris argues, in contrast to the prevailing view of the demilitarization ethos during the interwar period—and this is borne out by David Edgerton's *Warfare State*—invention and innovation in tank warfare remained key to British military thinking, so much so that Britain retained its superiority in the design of tanks, the training of tank crews and the strategic use of tanks in warfare. The Experimental Mechanical Force was formed in 1927, comprising tanks, other armoured vehicles and artillery. Its motto was one that could ultimately summarize the role of the flail tank during the Second World War: "No advance without security". In the lean years of military spending following the Great Depression and in the period leading up to the Second World War, a key proponent of the importance of the use of tanks in ground warfare was the commander

[12] "Australia to the Fore: Invention of the War Tank", *The Register*, 12 February 1920, https://trove.nla.gov.au/newspaper/article/65020239, accessed 18 December 2023.

[13] Tom Shachtman, *Laboratory Warriors: How Allied Science and Technology Tipped the Balance in World War II* (New York, Perennial, 2003), pp96, 98–99, 174–175.

[14] Harris, *Men, Ideas and Tanks*, p197.

of the newly formed Tank Brigade, Percy Hobart.[15] It was the adaptation of tanks termed "Hobart's Funnies" that would embody the spirit of innovation and adaptation to warfare that characterized the development of the tank. But the Second World War would also rekindle controversy over invention.

This essay addresses the invention of the flail tank during the Second World War by South African Abraham Stefanus Jacobus Du Toit, illustrating the sometimes challenging path of invention—and the inventor. The flail tank—and the controversy over its invention—provides some insight into the relationship between the "eccentricity" of invention within the context of industrialized warfare as well as the individual aspirations of the inventor (particularly one who is a member of the armed forces) set amidst the notion of service and duty to the war effort. Finally, woven into this case study, is the relationship between South Africa—a Dominion—and Britain as it relates to technology: its development, its production and its association with techno-nationalism.[16] As David Edgerton employs the term, techno-nationalism is the relationship between technology and the nation; it is through the lens of the nation that one can analyse the use of technology as it is the nation that makes possible the development of technology and its utilization. And it is the nation's invention and adoption of technology that confers a sense of status as "modern" or "developed". Yet, as Edgerton argues, the nation is not a self-contained entity and various extra-national, ideological and socio-economic factors come into play when considering techno-nationalism.[17] In this chapter, additional factors include a global war and Britain's historically dominant position in relation to the former colonies, one of which was South Africa. As Bill Freund shows, the outbreak of the War and Smuts's resumption of the role of Prime Minister marked the movement towards the "developmental state": the tremendous expansion in industry and the increasing role of the state in economic and industrial

[15] Refer to David Edgerton, *Warfare State: Britain, 1920–1970*, (Cambridge, Cambridge University Press, 2006) and Harris, *Men, Ideas and Tanks*, pp202, 217, 244.

[16] Edgerton, *Warfare State*, p104.

[17] Refer to David Edgerton, "The Contradictions of Techno-Nationalism and Techno-Globalism: A Historical Perspective", *New Global Studies*, Vol 1, No 1, 2007, pp1–32.

development. There was a focus too on technology and the machinery that would promote development.[18]

Similarly—but with a focus on the British setting—this study draws upon Edgerton's *Warfare State* that contextualizes the invention and production of the tank within the socio-economic and political environment of interwar and wartime Britain. Paul Kennedy provides valuable insight into the key moments of the War from 1942 that would ultimately culminate in an Allied victory, highlighting the significance of the use and adaptation of technology. In *Men, Ideas and Tanks*, Harris has an in-depth discussion of the development and perceived role of tanks in military strategy from the First World War and during the interwar period. The primary material used in this case study is drawn largely from the Department of Defence Archives that details the claims put before the Royal Commission.

The Flail Tank

The Royal Commission on Awards to Inventors was re-established in 1946 under R.G. Lloyd, a patent lawyer. Its board comprised members of the scientific and engineering fields with experts called in on a case-to-case basis.[19] By January 1947—the year in which the flail tank case was to be heard—the Commission had already received approximately fifty applications regarding wartime inventions by members of the Allied forces. In addition to the tank, the Commission had the onerous task of considering the validity of claims for "plastic armour for use on ships, amphibious tanks, the Bailey Bridge" and the intriguingly named "sticky bomb". Each claim required exhaustive investigation due to their "[raising of] complicated scientific and legal issues", subsequent to which the Commission would hold a hearing at its headquarters in Somerset House in London. Along with valid claims, the Commission had to contend also with the

[18] Bill Freund, *Twentieth-South Africa: A Developmental History* (Cambridge, Cambridge University Press, 2019), pp64–65.

[19] "Royal Commission on Awards to Inventors", *Nature*, No. 4013, 28 September 1946, p443.

frivolous from "inventors of strange and impractical schemes" coupled with vociferous demands for very generous compensation.[20]

In the case of the flail tank, the Commission was faced with two significant rival claims—the first being that of Major A.S.J. Du Toit. Du Toit was a Transport Officer and part of the Second Brigade. In October 1940 he was sent overseas, first serving in East Africa and then in North Africa. It was while experiencing the desert warfare conditions of the latter that Du Toit claimed to "[start giving] serious thought to the big problem of minefields, wire entanglements and tank traps". The tank would become his means of surmounting these obstacles "with a minimum sacrifice of life and material". As he recovered from a bout of dysentery in August 1941, he attended a lecture on the challenges of waging war in the desert and, in particular, the prevalence of minefields. At the end of the lecture, those attending were asked to think of possible solutions to the problem.[21]

Du Toit subsequently met with Brigadier Buchanan in charge of the South African base at Helwan in Egypt and drew up a sketch of his proposed "anti-mine flail device". He was then sent to the General Headquarters (GHQ) in Cairo to present his proposal and requested to return to South Africa to begin preliminary testing on the feasibility of the device.[22] According to a report submitted by GHQ to Brigadier Hansen, the Union Defence Force (UDF) Technical Liaison Officer, Du Toit's proposal was considered as "at first sight a mad idea" but with a "10 to 1 chance" of being feasible. Du Toit, on the other hand, was described as "enthusiastic" suggesting, unsurprisingly, less of a sense of reservation on his part. Hansen was requested to begin early testing of the "basic principles of the drum" with Du Toit being tasked to return to South Africa to elaborate on his vision.[23]

[20] "Royal Commission on Awards to Inventors", January 1947, File Q(T) 19/234, Vol 1: Inventions—Destruction of Mine Fields by Major A.S.J. Du Toit, KMG, Group 27, Box 143, DOD.

[21] Major A.S.J. Du Toit, "Affidavit", 7 November 1944, File Q(T) 19/234, Vol 1: Inventions—Destruction of Mine Fields by Major A.S.J. Du Toit, KMG, Group 27, Box 143, DOD.

[22] Major A.S.J. Du Toit, "Affidavit".

[23] Letter by Lance Robertson, GHQ, Cairo to Brigadier O. Hansen, Technical Liaison Officer, UDF, 21 August 1941, File Q(T) 19/234, Vol 1: Inventions—Destruction of Mine Fields by Major A.S.J. Du Toit, KMG, Group 27, Box 143, DOD.

On his arrival in South Africa in August 1941, Du Toit met with Hansen and requested ambitious testing of his proposal. Hansen initially refused due to the perceived difficulties in obtaining funding but, with a still "enthusiastic" Du Toit's persistence, a "[crude] model" was developed at the Technical Services Corps Base Workshop in Cullinan. To describe the model as "crude" was an understatement—chains had been attached to an oil drum mounted on the back wheels of a truck. Yet, it worked and as Hansen described it, "It successfully exploded the detonators of land mines and also destroyed wire entanglements". A now convinced Hansen believed it would be invaluable to the war. He then ordered the cessation of testing in order to maintain secrecy and, along with Du Toit, flew back to Cairo.[24]

Du Toit left preliminary sketches at GHQ in Cairo—which were to form the basis of the Scorpion tank—and was subsequently instructed to fly to England. Here he worked with representatives of the Ministry of Supply (MoS) to develop the device at the A.E.C. works in Southhall—an initial civilian site used for the building of buses.[25] Testing at Southall was far more ambitious than the rudimentary testing that had taken place in South Africa. The device—at this point known as the "Baron"—was tested on two different Matilda tank models, the Mark I and the Mark II. A rotating drum was mounted on metal arms that could be raised or lowered hydraulically. It was powered by an engine that was mounted to the side of the tank and attempts were made to prevent the overheating of the engine. As it stood, the device was still unwieldy. It also required two operators—the commander of the tank was to raise and lower the metal arms as required while giving the gunner instructions regarding the changing of gears. Key to the device—and its defining feature—were the chains and a chain was mounted on either end of the rotor with their movement "staggered" so as to ensure maximum flailing efficiency. A number of methods were attempted to determine ideal chain length and mounting: longer chains were likely to get tangled, shorter chains

[24] C.J. Hansen, "Flail Tank", 29 June 1944, File Q(T) 19/234, Vol 1: Inventions—Destruction of Mine Fields by Major A.S.J. Du Toit, KMG, Group 27, Box 143, DOD.

[25] Major A.S.J. Du Toit, "Affidavit".

were insufficiently flexible and could get damaged in the event of a mine blast.[26]

Related to the proposed efficiency of the flail device was the perceived layout of German minefields. According to a report by American military intelligence, the laying out of mines made use of natural and constructed obstacles in order to divert vehicles and troops into the minefield. While patterns and spacing varied, broad trends were detected in the layout of German mines. Tellermines—derived from the German for "plate"—used against tanks, for instance, were five paces apart when buried and ten paces apart when placed on the surface of the ground. If mines were placed close together, the setting off of a single mine could lead to the subsequent detonation of the others in proximity. If an open area had been mined in "haste", an area of about 580 square metres would likely contain about 12 mines at regular intervals. A deliberate minefield presented a more daunting task as each area would be "offset" by about 16 metres from the other, with approximately twice the number of mines working out to "a density of one mine per metre"—and all of these were subject to changing permutations.[27] The spacing of the chains would therefore need to account for these broad patterns and be adjusted accordingly.

The other feature of the tank would be its ability to cut barbed wire. The Mark I model was tested using the chains as a cutting tool—which proved unsuccessful. "Cross cutters" were however fitted to the Mark II model and proved effective with improvements suggested for subsequent models. The final feature was to be the tank's ability to deal with tank traps and here the device was successful with the rotor arms being used to dig up the dirt and "break" the tank trap.[28] During this testing period, Du Toit described the experiments as "most dangerous". Directly responsible for testing the device, Du Toit detonated numerous mines and, on

26 "The Du Toit 'Baron' Minesweeper: Report on Design, Test and Development—Mark I and Mark II", 18 September 1942, File Q(T) 19/234, Vol 1: Inventions—Destruction of Mine Fields by Major A.S.J. Du Toit, KMG, Group 27, Box 143, DOD.

27 US Military Intelligence Service, War Department, "German Coastal Defences", Special Series No 15, 15 June 1943, pp26–30.

28 "The Du Toit 'Baron' Minesweeper: Report on Design, Test and Development—Mark I and Mark II", 18 September 1942, File Q(T) 19/234, Vol 1: Inventions—Destruction of Mine Fields by Major A.S.J. Du Toit, KMG, Group 27, Box 143, DOD.

one occasion, a mine detonated prematurely against the tank, injuring his leg—an injury that would soon develop into arthritis.[29]

It was also at this point that Du Toit applied for a patent for the flail device—which had been refined by the testing process. In his British patent application filed in March 1942, Du Toit—who designated himself a "British subject"—described the tank as a means of "clearing" a minefield and thus permitting the safe passage of troops and vehicles. To do this the tank was equipped with a number of "attachments", somewhat akin to those found in a contemporary multitool. The first was "a beater device" designed to "beat" or "flail" the ground (hence the name "flail tank"). The device took the form of a drum mounted on a "U-shaped frame". The drum was itself designed to be "cage-like" rather than solid so as to dissipate the force of the explosion leaving it relatively undamaged. Attached to the drum were a series of "arms" to which were further attached chains. It was the latter that, propelled by the drum's rotation, would "flail" the ground, thereby setting off any mines that lurked just beneath the surface. The height of the drum could be adjusted in relation to changing topography by virtue of a "hydraulic mechanism" and Du Toit envisaged the drum and its frame as being attached to or detached from a tank as needed.[30]

In addition, the tank would also have attachments for the cutting of barbed wire or other impediments as well as destroying the "walls of tank traps". For the former, the drum would contain an initial device in the form of a cutting tool that Du Toit envisioned as being akin to a "grass-mower". While engaged in cutting, the chains would be rendered inert and safely "wrapped around the drum". These cutting tools or another similar device could also then be deployed in demolishing tank traps. Here, Du Toit notes their potential use in the traps that took the form of ditches dug into the ground in order to slow the progress of tanks and thus render them vulnerable to attack.[31] By the Second World War, tank traps and anti-tank measures were more sophisticated than they had been in the preceding conflict. The Germans made use of "dragon's teeth"

[29] Major A.S.J. Du Toit, "Affidavit".

[30] A.S.J. Du Toit, "Improved Apparatus for Clearing Land Mines or Minefields", Patent Specification 573, 753, 24 March 1942 (London, Patent Office, 1946), pp1–2, https://worldwide.espacenet.com/patent/search/family/009767438/publication/GB573753A?q=pn%3DGB573753A, accessed 15 December 2023.

[31] Du Toit, "Improved Apparatus for Clearing Land Mines or Minefields", p2.

which were rows of "concrete pedestals cast monolithically on a common base". The "teeth" varied in size, reaching up to 1.8 metres high and were designed to impede the progress of a tank or to "[belly] it". For the flail tank however, Du Toit envisaged it being able to contend with what more accurately could be termed an "anti-tank ditch". These ditches were dug into the ground and could be lined with earth, concrete or brick or even "steel rails", ranged from six to twelve metres wide and were often "laid in a zig zag course". Anti-tank ditches were often placed before "dragon's teeth" tank traps, creating a huge challenge for the safe passage of tanks.[32] Du Toit's flail tank however would only be effective against the more modest earth-lined anti-tank ditches.

Upon the successful conclusion of testing in December 1942, Du Toit "was informed that the device fulfilled all…possible expectations i.e. mine clearing, wire cutting and earth digging…and that it would be produced on a large scale and that it would be kept secret for the purpose of invasion".[33] This would have repercussions for his patent application and the credit he would receive. His work complete, he returned to South Africa.[34]

"The Man Who Invented the Flail Tank"

In the files related to the invention of the flail tank held at the Department of Defence Archives in Pretoria is an article from the London-based journal *South Africa*, appearing in August 1944. It is from a column titled "A South African's Diary" and appears to be a letter to the editor penned by a South African living in Bishopsgate in London. The letter details a description of the flail tank with the chains described as "a fearsome cat-o'-nine tails", the image of the whip for which the contraption was named. The author perceived the tank to be "a simple, crude, almost Heath Robinson idea", yet an effective one.[35] Robinson was an illustrator and cartoonist who acquired renown for his depiction of overly complicated gadgetry that was both eccentric yet utterly rational, "a world of

[32] US Military Intelligence Service, "German Coastal Defences", pp40–41.

[33] Major A.S.J. Du Toit, "Affidavit".

[34] Major A.S.J. Du Toit, "Affidavit".

[35] "The Man Who Invented the Tank Flail", *South Africa* 19 August 1944, p122, File Q(T) 19/234, Vol 1: Inventions—Destruction of Mine Fields by Major A.S.J. Du Toit, KMG, Group 27, Box 143, DOD.

pulleys, wires, strings and cogs powered by steaming kettles". Such inventions would be described as "a Heath Robinson contraption" by the end of the First World War and would find their way into the cartoons that he drew during the Wars. Robinson's work was marked by a sense of whimsy and the ludicrous nature of the technology of war that did much to obviate its association with violence and death: the use of laughing rather than mustard gas, flying tanks, narrow tanks to negotiate tight spaces and the use of immense cannons to propel the Germans to the English coast.[36] The odd appearance of the flail tank and the jury-rigged adaptations that marked its beginning made the analogy an apt one. Yet, while Robinson's illustrations exemplified the eccentric British inventor, the letter writer emphasized the South African origins of the invention, chastising the UDF for its reluctance in making known the inventor of the tank.[37] Yet the road to recognition for Du Toit was a difficult one—and one that was not necessarily due to the recalcitrance of the UDF.

There were two factors responsible for this lack of recognition. The first was the patent itself. As mentioned earlier, during the testing of the device in Britain, Du Toit applied for the British patent based on advice he claimed to have received from the Secretary of Defence suggesting that he should consult a "patent agent and file application for a patent". Should his device then be used he would be entitled to "compensation".[38] Yet, the advice obtained from the South African Department of Defence would be subordinate to the demands of the War as determined by the British MoS. For the UDF, any consideration of a financial reward for Du Toit would only come after the War had ended—and when the efficacy of the flail tank could be proven—and would further depend on Du Toit's "agreements" with the British.[39]

Du Toit's patent agents in London were Marks and Clerk and he submitted a provisional patent application on 10 March 1942 with the "Final Specifications and Drawings" submitted on 10 December of that

[36] Richard Holledge, "The Absurdities of War: How Heath Robinson Eschewed Jingoism", *The New European*, 18 August 2022, https://www.theneweuropean.co.uk/the-absurdities-of-war/, accessed 2 January 2024.

[37] "The Man Who Invented the Tank Flail".

[38] Du Toit, "Affidavit".

[39] Cecil D. Trollip for Secretary for Defence, "Invention: Destruction of Minefields – Major A.S.J. Du Toit", 27 April 1942, File Q(T) 19/234, Vol 1: Inventions—Destruction of Mine Fields by Major A.S.J. Du Toit, KMG, Group 27, Box 143, DOD.

year. Upon his return to South Africa, he requested that his South African patent agents, Kisch and Coy, use the British application to apply for the patent in Canada and the United States. Kisch and Coy were, however, unable to get permission to do so due to what the British Patent Office claimed to be "secrecy measures"—of which Du Toit professed ignorance.[40] Concerned that he had flouted British regulations, yet also aware that he had a small window in which to register his patent in North America, Du Toit "[pleaded] ignorance and the misdirection of his patent agents".[41] Marks and Clerk took exception to Du Toit's claim of "misdirection" and their perception was upheld by the MoS—who attributed the incident to "misunderstanding" on the part of Du Toit.[42] However, the MoS had no objection to Du Toit filing the American patent provided that they determined the American patent lawyer; that all relevant documents required for the filing be submitted by them only; and that Du Toit could only "disclose the invention" if given written permission to do so by the MoS and was, further, to ensure all measures "to prevent unauthorized disclosure".[43] For the Dominions—Australia, Canada, New Zealand and South Africa—no patent agents were to be appointed, the approval of the application could be "postponed indefinitely" based on the requirements of confidentiality and the MoS remained in "complete control" of the application process.[44] From the South African perspective, there could be no challenge to the notion of "secrecy" as according to a War Measure on patents issued in 1940, if an invention was determined to be "secret" when abroad, it would remain so in South Africa until "the

[40] Du Toit, "Affidavit".

[41] Telegram 3/327/10, 22 May 1943, File Q(T) 19/234, Vol 1: Inventions—Destruction of Mine Fields by Major A.S.J. Du Toit, KMG, Group 27, Box 143, DOD.

[42] G.W. Carlton, Ministry of Supply, 16 June 1943, File Q(T) 19/234, Vol 1: Inventions—Destruction of Mine Fields by Major A.S.J. Du Toit, KMG, Group 27, Box 143, DOD.

[43] "British Patent Application No", Ministry of Supply, File Q(T) 19/234, Vol 1: Inventions—Destruction of Mine Fields by Major A.S.J. Du Toit, KMG, Group 27, Box 143, DOD.

[44] Ministry of Supply, "Rules and Conditions for Filing Patent Applications in Australia, Canada, New Zealand and South Africa, Corresponding with British Patent Applications made Secret under Regulation 6(1) of the Defence Regulations 1939, upon the Advice of the Ministry of Supply", File Q(T) 19/234, Vol 1: Inventions—Destruction of Mine Fields by Major A.S.J. Du Toit, KMG, Group 27, Box 143, DOD.

overseas 'secrecy' is lifted".[45] Du Toit's American patent was eventually filed on 31 August 1945 (after the end of the War) and registered on 29 November 1949.[46] It was these "secrecy measures" implemented by the MoS that would account for the early lack of acknowledgement given to Du Toit—and pave the way for rival claimants.

The article appearing in *The South African* stated that the South African High Commissioner to the United Kingdom, Colonel Deneys Reitz, believed that the flail tank was the brainchild of the son of a Senator Conroy[47] and, upon reading the article, Du Toit came to the conclusion that such an endorsement from Reitz was "prejudicial" to his own case for recognition. Ironically, it was the very success of his invention that had contributed to the lack of acknowledgement. According to Du Toit, his invention had been placed on the "War Office's 'Secret List'" and other than his patent application, he therefore took no further measures to assert his claim to the invention. It was when confronted with the article in *The South African* that accredited another inventor along with his belief that the device was no longer on the "secret list", that Du Toit now put forth his claim. Along with his affidavit, were supporting documents provided by figures involved in the development of the device.[48] In his report on the development of the flail tank, for instance, Brigadier Hansen, the UDF Technical Liaison Officer who was involved in the very rudimentary testing of the device conducted in South Africa, felt that the lack of acknowledgement given to Du Toit for his role in inventing the flail device was an "injustice" that needed to be redressed.[49]

Du Toit had to contend with a more serious rival claim than that of the younger Conroy. In a statement made to the UDF in 1944, Harold Mill-Colman claimed to have been inspired to invent the flail tank during the fall of Tobruk in June 1942.

[45] "Re Memorandum Re War Measures", 13 October 1944, File Q(T) 19/234, Vol 1: Inventions—Destruction of Mine Fields by Major A.S.J. Du Toit, KMG, Group 27, Box 143, DOD.

[46] A.S.J. Du Toit, "Apparatus for Clearing Land Mines or Minefields", 2, 489, 564, United States Patent Office, 29 November 1949.

[47] "The Man Who Invented the Tank Flail".

[48] Major A.S.J. Du Toit, "Re: Claim of No 86663(V) Major A.S.J. Du Toit as Inventor of 'Anti-Mine Flail Device'", 10 November 1944, File Q(T) 19/234, Vol 1: Inventions—Destruction of Mine Fields by Major A.S.J. Du Toit, KMG, Group 27, Box 143, DOD.

[49] Hansen, "Flail Tank".

In the midst of what was a humiliating outcome for South African forces, Mill-Colman noticed "an Army vehicle with a length of barbed wire entangled in the wheel". It evoked the image of a "threshing unit" to be used "as a new weapon of war…against MG posts [and] as a morale breaker". At that point "a scheme [began] formulating in his mind". It would be during the first battle of El Alamein that the "scheme" began to take shape as Allied forces were confronted with extensive enemy minefields.[50]

Yet the UDF had also begun looking at other anti-mine measures, albeit on a less ambitious scale than the flail tank. The W.A.C. Anti-Tank Mine Detector epitomized harmonious co-operation between civilians and the military. It was developed by Oscar Weiss, a geophysicist in Johannesburg, and was based on an idea by Lt Col A.G. Clark of the 3rd Division Group and its development funded by another civilian, P.M. Anderson. Weiss would subsequently go on cede his patent rights to the UDF.[51] Weiss's mine detector would cost the UDF little more than £150 to develop and produce, and its specifications would be determined by the capabilities of the human body: a total weight of not greater than eleven kilograms, and a length of approximately 1.8 metres, giving it a range of twice that. Designed to be worn by the individual for a long period of time under dangerous conditions, it was to be attached to a harness and would also come equipped with a helmet specially designed to incorporate the wearing of earphones. Finally, it needed to be sufficiently "robust" to withstand battlefield conditions: its transportation to front lines and the possibility of its user coming under enemy fire and having to take evasive action.[52]

Mine detectors were in effect metal detectors and explicit instructions were provided by Lt. Col Clark regarding the operation of the W.A.C. Mine Detector. The operator was not to carry any article made of metal:

[50] Lt. Col. Mill-Colman, "Scorpion MK I", 30 September 1944, File Q(T) 19/234, Vol 1: Inventions—Destruction of Mine Fields by Major A.S.J. Du Toit, KMG, Group 27, Box 143, DOD.

[51] Letter by Lt. Col. A.G. Clark, "3 Div W.A.C. A.T. Mine Detector", File DTS 19/238: Inventions—General Anti-Tank Invention, Invention by Mr Weiss, KMG, Group 27, Box 143, DOD.

[52] Letter by Brigadier General Cecil D. Trollip, Director General: Technical Services, "Anti-Tank Mine Detector", 1 May 1942, File DTS 19/238: Inventions—General Anti-Tank Invention, Invention by Mr Weiss, KMG, Group 27, Box 143, DOD.

a knife, wristwatch and even the buttons of his shirt were not to be "attached by split steel rings". The harness was to be donned with the detector held in the right hand and switched on. If the detector produced "a sharp yap yap like noise" when a metal object was placed close to it, it was ready for use. It could then be swept back and forth over the prospective minefield, detecting the presence of mines (and other metal objects). Once the mine was located, its position could be marked for it to be disarmed.[53] An initial order for twelve W.A.C. Mine Detectors was made by the UDF "as a matter of urgency" with the aim of them being sent to the Middle Eastern theatre of war[54]—however the device would be supplanted by a competitor.

On 12 January 1942, both the W.A.C. Mine Detector and another detector developed by the Director General Technical Services (DGTS) were tested at the Military College, Voortrekkerhoogte. The ground was level, contained significant levels of iron and was slightly wet, enabling conductivity. The mines were buried approximately 5 centimetres deep and the earth surrounding them was further dampened. These formed part of two courses, each about a metre wide and 25 metres long, along which the operators tested the mine detecting equipment. The DGTS version completed the course in less than fifteen minutes, detecting all the buried mines. Its W.A.C. counterpart took three minutes longer, failed to detect one of the mines and thus performed more poorly than it had done in the past due to a possible "loose connection" giving "erratic results". Ultimately the DGTS version was recommended due to its convenience: it weighed less, required fewer batteries and less maintenance and was of simpler design. Despite the earlier and strong test performance of the W.A.C. model, it was recommended that the DGTS mine detector be put into practice.[55] Yet, just as these mine detectors demonstrated

[53] Letter by Lt. Col. A.G. Clark, "3 Div W.A.C. A.T. Mine Detector", 16 December 1941, File DTS 19/238: Inventions—General Anti-Tank Invention, Invention by Mr Weiss, KMG, Group 27, Box 143, DOD.

[54] K.R. van der Spuy, DGTS, "W.A.C. Anti-Tank Mine Detector", 20 December 1941, File DTS 19/238: Inventions—General Anti-Tank Invention, Invention by Mr Weiss, KMG, Group 27, Box 143, DOD.

[55] A.R. Hamer-Philip, "Detectors Mine A.T. Types A and B. Comparative Test", 13 January 1942, File DTS 19/238: Inventions—General Anti-Tank Invention, Invention by Mr Weiss, KMG, Group 27, Box 143, DOD.

their efficacy in countering Axis mine fields, their success also highlighted their weakness. To counter the detectors, Clark recommended the encasing of mines in manganese steel, an amalgam of manganese, carbon, sulphur, phosphorus and silicon, thus rendering them non-metallic and hence undetectable.[56] The incident echoed that of the Torpedo Mine B (TMB) used by the Germans against British shipping during the Second World War. The British acquired a torpedo had their engineers develop successful countermeasures and then improved their own magnetic mines to be used against German shipping. It is an illustration of the context in which war technology develops, its evolving and adaptable nature—and the ways in which a defensive measure can become an offensive weapon.[57]

Mine detectors, however, were only the first part of the process of desert warfare in 1942. The second part was the removal of the threat posed by these mines so as to allow the unimpeded advance of Allied forces. While the UDF had the resources to test the more modest anti-mine detectors, the same could not be said for the flail tank (and this was why Du Toit was dispatched to Britain to oversee the testing of the device). Mill-Colman's idea would be far bolder, and he was thus reticent as he initially presented it—along with a pencil drawing—to Brigadier J.K. Ray, CE 30 Corps. The scheme required specialized equipment, skilled workers and the capability of mass production—along with the initial capital to develop it. Clearly impressed, Ray subsequently allocated 21 Corps Field Park Company under Major Girling to work with Mill-Colman in developing the tank. Leslie Alfred Girling was 34 at the time of his enlistment and his initial posting was as a second lieutenant with the Engineering Corps at Premier Mine in May 1940, becoming second in command of the 21 Corps Field Park Company at the time the Scorpion was being developed. Ray also allowed the Company the use of any material required that could be obtained from Allied bases in Egypt. An engine and chassis were scrounged from an earlier abandoned attempt at a flail tank and these would be incorporated into the prototype. Initially named the "Durban MK I" in honour of Mill-Colman's origins, Ray renamed the tank the "Scorpion" with its subsequent biblical resonance: "And now whereas my father did lade you with a heavy yoke, I will add to

[56] Lt. Col G.A. Clark, "Anti-Tank Mines", File DTS 19/238: Inventions—General Anti-Tank Invention, Invention by Mr Weiss, KMG, Group 27, Box 143, DOD.

[57] Shachtman, *Laboratory Warriors*, pp98–99.

your yoke: my father hath chastised you with whips, but I will chastise you with scorpions".[58]

Upon the successful completion of a number of tests, an initial order was made for 24 tanks—based on drawings made by Major Girling—and personnel were then trained to operate them, readying them for use at the Second Battle of El Alamein.[59]

Girling's statement supported that of Mill-Colman. Also made to the UDF in 1944, he bemoaned the dispute over the development of the Scorpion, especially after its lauded performance at El Alamein. Girling also emphasized the personal accolades he had received for his role in its development, including an M.B.E. on the recommendation of Brigadier Ray.[60] His citation stated in part:

> Major Girling has between 6 Aug and 15 Sep 42 played a very large part in the design and construction of the mine-destroying device called Scorpion Mk 1., and it has largely been through Major Girling's energy and determination that this device has been successfully developed in the field.[61]

As with Mill-Colman, Girling's account began with the first battle of El Alamein where it became increasingly apparent for the need "eliminate 'mine consciousness'…amongst Tank Crews" which suggests the increasing apprehension of tank crews facing Axis minefields. In addition to the detrimental psychological effect of traversing minefields, there was of course the need to open up mine-free passages to allow troops and equipment to pass. As a result, the Eighth Army created the Army School of Mine Clearance, and it was at this point that Mill-Colman put forward his notion "of a machine using chains as a flail to beat the ground and

[58] Lt. Col Mill-Colman, "Scorpion Mk. I", 30 September 1944, p2, File Q(T) 19/234, Vol 1: Inventions—Destruction of Mine Fields by Major A.S.J. Du Toit, KMG, Group 27, Box 143, DOD; 1 Kings Chapter 12, Verse 11 and "Girling, Leslie Alfred", Service Card No W.R.12/2064, 123668V, DOD.

[59] Lt. Col. Mill-Colman, "Scorpion MK I".

[60] Major Girling, "Subject: The Scorpion", 5 September 1944, File Q(T) 19/234, Vol 1: Inventions—Destruction of Mine Fields by Major A.S.J. Du Toit, KMG, Group 27, Box 143, DOD.

[61] Reproduced in Director of Engineers, "Scorpion (Anti-Mine Flail Device): Inventor's Claims", 19 January 1945, File Q(T) 19/234, Vol 1: Inventions—Destruction of Mine Fields by Major A.S.J. Du Toit, KMG, Group 27, Box 143, DOD.

so detonate any mines in their path". Interestingly enough, Girling could not account for the actual origins of the idea, suggesting an apocryphal story that it may have been Winston Churchill.[62]

Girling's statement echoed that of Mill-Colman in terms of the development of the tank—and the chains were affixed to an existing Matilda tank. Seated atop a truck and with the assistance of two draughtsmen, Girling was responsible for drawing up the sketches and then ensuring that the relevant parts were built to his specifications—a process that took just over a week. Tests were subsequently carried out and tank crews were then trained in its use. Although initially trepidatious due to "mine consciousness" as well as the intimidating appearance of the Scorpion—described as a "fearsome monster"—the efficacy of the tank in detonating mines while remaining relatively unscathed, soon converted the tank crews. While acknowledging various figures who had aided in its development, Girling concluded his report by categorically denying that any external assistance was received. With regard to Du Toit, he stated the following: "We believed that Du Toit (I do not know his title) was working on a similar type of machine in England, but not one atom of information was received from him".[63]

While supporting statements had been presented by UDF personnel favouring either the case of Du Toit or that of Mill-Colman, the UDF did not officially wish to be drawn into the fracas. A statement by the DGTS made clear that the UDF did not have a procedure in place regarding the recognition of the "rights to inventions" made by serving officers and further had "no interest in the invention".[64] According to the Chief of General Staff (CGS), the dispute over the origin of the invention was "a potentially contentious subject" and one in which it was "inadvisable" for the UDF to play an active role.[65] As a result, the UDF "left the matter to

[62] Major Girling, "Subject: The Scorpion", 5 September 1944, File Q(T) 19/234, Vol 1: Inventions—Destruction of Mine Fields by Major A.S.J. Du Toit, KMG, Group 27, Box 143, DOD.

[63] Girling, "Subject: The Scorpion".

[64] Letter by DGTS, 20 December 1945, File Q(T) 19/234, Vol 1: Inventions—Destruction of Mine Fields by Major A.S.J. Du Toit, KMG, Group 27, Box 143, DOD.

[65] GSO, CGS Section, "No 86663V Major A.S.J. Du Toit 'Anti-Mine Flail Device'", 9 August 1945, File Q(T) 19/234, Vol 1: Inventions—Destruction of Mine Fields by Major A.S.J. Du Toit, KMG, Group 27, Box 143, DOD.

the United Kingdom authority".[66] Given the subordinate role accorded to the UDF by the MoS over the registration of Du Toit's patent, this was hardly surprising.

Based on the various statements by the figures involved, a timeline was submitted by the Acting Quarter Master General of the UDF to the Principal Patents and Awards Officer at the Ministry of Supply, London. It also alluded to the availability of the supporting documentation to substantiate the timeline which was as follows:

a. Du Toit started with ideas in August 1941
b. Du Toit filed patent application in March 1942 in the United Kingdom
c. Mill-Colman conceived idea in June 1942
d. Mill-Colman worked on Scorpion independently from Du Toit[67]

As the case was heard before the Royal Commission for Awards to Inventors—and eighteen months after the dispute began—Du Toit was asked to make a progress report to Brigadier John Daniel in charge of the administration of UDF troops based in Britain. By this point, Du Toit—who was theoretically attached to the UDF Administrative Headquarters in London—was utterly consumed by the case, with little time for his official duties. He informed Daniel that, despite the intervening year and a half and the prohibitive cost incurred in terms of his legal fees—for which he would be accountable regardless of the outcome—"he [had] committed himself financially to such an extent that he [had] to see the matter through", and he anticipated that there would be an outcome by July 1947.[68] It was both a financial and a very human cost. And there were unexpected challenges.

Less than three weeks before an official recommendation was made regarding those who were involved in the development of the flail tank, a somewhat anticlimactic moment occurred when a letter was sent to the

[66] Letter by DGTS.

[67] W. Driver for Major-General, Acting Quartermaster General, "Flail Tanks: Lt. Col Mill-Colman", 1 June 1948, File Q(T) 19/234, Vol 1: Inventions—Destruction of Mine Fields by Major A.S.J. Du Toit, KMG, Group 27, Box 143, DOD.

[68] Brigadier John Daniel, "No 86663V WS. Major ASJ Du Toit", 16 April 1947, File Q(T) 19/234, Vol 1: Inventions—Destruction of Mine Fields by Major A.S.J. Du Toit, KMG, Group 27, Box 143, DOD.

Adjutant General in Pretoria by Corrie Michael Barendse who claimed to have been ordered to work on the flail tank project along with Du Toit in 1941. In his letter, Barendse—who was a "motor engineer" and had served as a second lieutenant at the Base Workshops in 1941—stated that Du Toit was "always absent", leaving him in the sole role of developing the tank, which was subsequently tested, with the successful "experiment" captured on film. He was then informed by his commanding officer that he would likely have to go to the United States or to England in order to demonstrate the device. Before such could occur however, Barendse became one of the 33,000 taken prisoner at Tobruk and would remain a prisoner-of-war for three years. Upon his release in 1945—and hearing of the controversy regarding the accreditation of the invention of the tank—he claimed that "whatever anyone has done towards the Flail Tank, it is only an improvement of my original invention".[69]

After investigating the validity of possible claimants, a final list was submitted by the South African High Commission to the Royal Commission on Awards to Inventors in August of 1948.[70] The list comprised sixteen candidates with a recommendation for remuneration based on their perceived contribution to the development of the tank. The amounts suggested ranged from £13,000 (Du Toit) to £100. Four candidates were deemed not to require any remuneration, and two candidates withdrew their applications. Barendse was not mentioned.[71] The Commission paid heed to the South African recommendation: ten men received an award for the development of the flail tank, five of them were South Africans.[72] Du Toit emerged with the lion's share of £13,000 followed by Mill-Colman with £2,000 and Girling and Norman Berry with £1,000 each.[73]

[69] Letter by C.M. Barendse, 22 July 1948, p2, File Q(T) 19/234, Vol 1: Inventions—Destruction of Mine Fields by Major A.S.J. Du Toit, KMG, Group 27, Box 143, DOD and "Barendse, Corrie Michael", Service Card No W.R.12/8091, 240423, DOD.

[70] "List of Awards Recommended to the Treasury in Connection with the Development of the Flail Tank", 10 August 1948, File Q(T) 19/234, Vol 1: Inventions—Destruction of Mine Fields by Major A.S.J. Du Toit, KMG, Group 27, Box 143, DOD.

[71] "List of Awards Recommended to the Treasury in Connection with the Development of the Flail Tank".

[72] "£20,000 Award for Inventors of Flail Tanks", *The Times*, 2 September 1948, https://trove.nla.gov.au/newspaper/article/78590436, accessed 15 December 2023.

[73] "List of Awards Recommended to the Treasury in Connection with the Development of the Flail Tank".

Cometh the Hour, Cometh the Tank…

Major Norman Berry had been dispatched to South Africa to observe the early tests with Du Toit's "threshing machine". While in North Africa, he subsequently began attempts to render the device practical for use by the Eighth Army. This would culminate in the use of the Scorpion during the Second Battle of El Alamein.[74] As Kennedy discusses, the Allied victory at El Alamein was largely due to defensive warfare and, in particular, the laying out of extensive minefields that presented an almost insurmountable obstacle to Erwin Rommel's advance and depleted his already low fuel reserves. Yet Rommel would also adapt the use of minefields against the Allied Eighth Army, and it is here that anti-mine devices came to the fore. These included the "handheld acoustic mine detector" invented by Josef Kosacki of the British Polish army, and variants of which were developed and tested in South Africa as discussed above. And the other was the Scorpion flail tank that could clear a path through the German minefields.[75]

Adaptations to the Scorpion led to the use of a Sherman rather than Matilda tank which would become known as the Sherman Crab.[76] This adaptation was not unexpected. As Edgerton states, with the exception of the Vickers tanks, British tanks during the interwar period were not necessarily of a good standard and the preference was for American Grant and Sherman tanks.[77] As an improvement on Du Toit's original design, the rotating drum containing the flails did not require a separate engine but was powered directly from the engine of the tank itself. The Sherman Crab become the standard flail tank for the 79th Armoured Division under tank specialist Percy Hobart that was to play such a key role on D-Day.[78]

The great Allied offensive of the Second World War was Operation Overlord. Commencing on 6 June 1944 (D-Day) a combined Allied invasion landed on the beaches of Normandy at five points codenamed

[74] Richard Doherty, *Hobart's 79th Armoured Division at War: Invention, Innovation and Inspiration* (Barnsley, Pen and Sword Books Ltd, 2011), Chapter 4.

[75] Kennedy, *Engineers of Victory*, pp161–163.

[76] Doherty, *Hobart's 79th Armoured Division at War*, Chapter 4.

[77] Edgerton, *Warfare State*, p81. As the War drew to a close however, the British would develop the more formidable Centurion tank.

[78] Doherty, *Hobart's 79th Armoured Division at War*, Chapter 4.

Sword, Juno, Gold, Omaha and Utah. This combined Allied invasion force was the largest in history—submarines and minesweepers, vessels developed for amphibious landings, more than a thousand ships, upwards of 300,000 troops and aerial support comprising almost 12,000 planes, including 5,600 fighters. Also evident were 2,000 tanks.[79] They were confronted with beaches strewn with barbed wire and mines and facing German gun batteries that were part of the Atlantic Wall—a series of defensive fortifications designed by Rommel running along the coast and designed to thwart invasion. The tank divisions in particular had spent the time prior to D-Day preparing to contend with the challenges posed by the Wall: "enormous sections of walls of different heights and thicknesses, minefields of anti-tank and anti-personnel mines, wire entanglements, anti-tank ditches, all of which we practiced blowing up, cutting through or crossing".[80]

"Hobart's Funnies" were emblematic of what Paul Kennedy refers to as "problem solving"—the adaptation and improvement of technology to achieve strategic goals. American Sherman and British Churchill tanks were adapted for the D-Day landings—there were amphibious tanks, tanks equipped with flamethrowers, tanks with their own bridges to negotiate tank traps and tanks that simply allowed the passage of other tanks.[81] And key to Operation Overlord were the flail tanks, the anti-mine tanks that paved the way for the storming of the beaches: "They drove off the beach flailing...They flailed straight up to the dunes, then turned right flailing and then flailed back to the high-water mark".[82] Three weeks after D-Day, the South African newspaper, *The Rand Daily Mail*, depicted a picture of the flail tank in action, the chains a blur as they beat the earth. The excerpt was titled, "South African Invention Used on Normandy Battlefield".[83] Du Toit was not mentioned—his role as inventor would only come to the fore with the award by the Royal Commission on Awards to Inventors four years later.

[79] Refer to Kennedy, *Engineers of Victory*, Chapter 4.

[80] Doherty, *Hobart's 79th Armoured Division at War*, Chapter 5.

[81] Kennedy, *Engineers of Victory*, pp370–371, 270.

[82] Quoted in Kennedy, *Engineers of Victory*, p271.

[83] "South African Invention Used on Normandy Battlefield", *Rand Daily Mail*, 30 June 1944.

From its first appearance on the Somme, the tank typified the mechanized, alienating and modern face of industrial warfare. Associated—with some ambiguity—with British ingenuity, its development and production during the interwar period challenged the image of a state ill-equipped to deal with modernization. Yet its invention is a human story, and the tank simultaneously upheld a specific vision of British technological development as the manifestation of the eccentric genius of the lone inventor, a weapon improvised for the demands of war. The flail tank exemplified this image of British wartime weaponry as defensive rather than offensive in contrast to the clinical killing machines associated with German bellicosity.[84]

The invention of the flail tank and the disputes arising from it demonstrated too techno-nationalism within the context of war and empire. Despite the inventor (and his rival claimant) being South African, the lack of resources meant that the practical development of the tank was undertaken with British expertise and resources. As such, while ostensibly suggestive of Allied collaboration, the British Dominion of South Africa was relegated to a supportive role; the Ministry of Supply retained control over the invention and its dissemination, ultimately pitting the individual ambitions of the inventor against the nationalism associated with service, security and the Allied war effort.

[84] Refer to Edgerton, *Warfare State,* pp273–276.

CHAPTER 5

A Tale of Two Hospitals: Reconstructive Surgery and Race

"Our wounded had call upon surgical skill from the whole Anglo-Saxon race".[1] Plastic surgery pioneer Harold Delf Gillies was referring to the formidable assemblage of medical professionals drawn from across the British Empire that had gathered at the Queen Mary's Hospital in Sidcup and were key to the development of modern plastic and reconstructive surgery during the First World War. Modern plastic surgery, as evident at Sidcup, was a response to and contextualized by the injuries incurred on the Western Front. The great range of injury—usually due to projectile weapons such as bullets and shells—provided the impetus for the development of surgical techniques to rectify facial—and bodily—damage that was recorded in Gillies' *Plastic Surgery of the Face*.

Gillies' book was a landmark in plastic surgery and employed a singular visual pattern in the depiction of facial injury. A photograph was taken of the patient's face, focusing on the injury prior to surgery. This was then followed by drawings by artist and surgeon, Henry Tonks, that leant graphic detail to the depiction of the injury. A final photograph was taken

[1] Quoted in John D. Holmes, "Development of Plastic Surgery", *War Surgery, 1914–1918,* Thomas Scotland and Steven Heys, (eds) (Solihull, Helion and Company, 2012), p265.

S. Chetty, *Science and Technology in South Africa, 1939-1946*, Palgrave Studies in the History of Science and Technology,
https://doi.org/10.1007/978-3-032-05084-7_5

of the face after surgery and recuperation, demonstrating a marked difference in appearance that signalled a return to "normality". *Plastic Surgery of the Face* also detailed the methods of reconstructive and plastic surgery. Forms of intervention by plastic surgeons included the use of skin flaps to cover scarring, "fat flaps" to flesh out gaping wounds as well as prosthetics to substitute for missing bone, cartilage and teeth.[2] In the midst of the ravages of trench warfare, much had been expected of plastic surgeons confronted with the novel and, as Gillies recollected later: "Unlike the student of today, who is weaned on small scar excisions and graduates to harelips, we were suddenly asked to produce half a face".[3]

Gillies' major surgical contribution was the tube pedicle: the use of a graft obtained from the chest or the forehead which would be used on severe facial damage. The graft remained partly attached to its point of origin to ensure blood flow. One of the complement of surgeons attached to Gillies' first plastic surgery unit at the Cambridge Military Hospital in Aldershot and then Sidcup was John Law Aymard, described by Gillies as "a South African practitioner with rather an odd history".[4] In an article appearing in *The Lancet* in December 1917, Aymard described his "new method" of reconstructing the nose which "consists in taking a long pedicle flap with its base near one clavicle, and forming the nose upon the chest, on the opposite side, keeping the head slightly bent upon the chest for 14 days". The input of a sculptor as well as Henry Tonks was required to render an aesthetic result. Accompanying his lengthy description of the process, Aymard made use of rough sketches, as well as photographs and X-rays of patients. At the time of the publication of the article however, treatment was as yet incomplete due to Aymard having to return to South Africa. He considered the preliminary results significant however to bear publication.[5] Gillies, however, would claim to have

[2] H.D. Gillies, *Plastic Surgery of the Face* (London, Henry Frowde and Hodder and Stoughton, 1920), Chapter V, p48.

[3] Caroline Alexander, "Faces of War" in *Smithsonian Magazine*, February 2007, https://www.smithsonianmag.com/arts-culture/faces-of-war-145799854/, accessed 12 August 2020.

[4] Quoted in Andrew Bamji, *Faces from the Front: Harold Gillies, The Queen's Hospital, Sidcup and the Origins of Modern Plastic Surgery* (Solihull, Helion and Company Ltd, 2017), p49.

[5] J.L. Aymard, "Nasal Reconstruction with a note on Nature's Plastic Surgery", *The Lancet*, 15 December 1917, pp888–891.

used the revolutionary method before Aymard, leading to an acrimonious exchange via *The Lancet* with Aymard claiming primacy as late as 1935.[6]

In October 1917, an exhibition was held at the Hunterian Museum "of the Army collection of war specimens". The reconstructive surgery exhibits were on display in Room III and included plaster casts, wax models, photographs (including colour photographs) and X-rays of facial injury, demonstrating to the public the types of facial injury, the methods of treatment by Gillies and Aymard and the results. While the War was still ongoing, the reason for the exhibition was articulated in the catalogue accompanying the exhibition: the accumulation and sharing of medical expertise derived from the experiences of war, with the hope that all that had been learned would be "[consolidated]...for the use of a future time and generation".[7]

With its origins on the Western Front, modern plastic and reconstructive surgery demonstrated from the outset two key features: a collaborative—albeit occasionally rancorous—effort between the surgeons drawn from the Dominions that nonetheless placed Britain at the centre of development and surgical expertise; and innovation derived from the casualties of war who were inevitably white and male. The specialization's origins can therefore be contextualized by a particular racial (and gendered) development, both in terms of medical professionals as well as those treated. This, however, is a feature inherent in western medicine.

In approximately 1490, Leonardo da Vinci drew the Vitruvian Man in pen and ink. One of his most recognizable drawings, it is part of a series depicting the "divine proportions" of the human face and form that so intrigued the artist, inventor, scientist and anatomist who came to define the Renaissance. Influenced by the writings on "harmonious proportions" by the Roman, Vitruvius, fifteen hundred years earlier, Da Vinci's drawing also symbolized the resurgence of classical Greek and Roman influences that were a hallmark of the Renaissance. Using the male body as the archetype for the human form, the Vitruvian Man depicts a nude figure in two poses designed to show the ratio between limbs, body and digits. More than a mathematical ratio, it is also a depiction of the ideal male

[6] J.L. Aymard, "Nasal Reconstruction: The Tubed Pedicle Flap", *The Lancet*, 27 April 1935, p1018.

[7] "The Army Collection of War Specimens: The Opening of the Exhibition", *The Lancet*, 20 October 1917, pp620–621.

form emphasizing musculature and physicality.[8] Da Vinci's creation was emblematic of the humanism of the Renaissance with its emphasis on the perfectibility of the individual. Physical perfection was thus manifest in the Vitruvian Man.

Drawing upon the work of Michel Foucault however, Urla and Terry make the point that bodies are not neutral sites but are contextually constructed and imbued with the exercise of power—or lack thereof. The ways in which the body is represented in both medicine and culturally relates therefore to societal and other distinctions. It is difference that lies at the heart of representing the body with the unspoken "ideal human body" that of the "robust, European, heterosexual gentleman". This ideal body is the norm against which other bodies are measured, and found wanting; bodies marked by their race, class, gender or state of health. This is compounded by the scientific notion of the material nature of the body, its very corporeality giving it the status of "truth" while simultaneously silencing its constructed nature.[9]

Within South Africa, the body and race took on a particular cogence, in a context of segregation as well as debates about human origins. In May 1938, Robert Broom made an astonishing claim in a lecture presented at the University of the Witwatersrand:

> Evolution is finished and man in the future, therefore, will probably be very little different physically from what he is today. But the evolution of his brain will continue. The superman of several thousand generations to come, I think we can assume quite safely, will be similar physically to man as we know him today, but he will have a much higher type of brain…[10]

Broom's view of current humanity—at least as it related to physical development—as the apex of evolution was controversial as was his belief that evolution, as it applied to the entire animal kingdom was now complete. It held to Broom's view of evolution as progressive and possibly guided

[8] Charles Nicholl, *Leonardo da Vinci: The Flights of the Mind* (London, Penguin Books, 2007), pp245–247.

[9] Jacqueline Urla and Jennifer Terry, "Introduction: Mapping Embodied Deviance", *Deviant Bodies*, Jennifer Terry and Jacqueline Urla, eds. (Bloomington and Indianapolis, Indiana University Press, 1995), pp3–5.

[10] "Stature May Halt, Brain Will Increase: Dr Broom Visualises Superman", *Rand Daily Mail*, 21 May 1938.

by the hand of the divine. Its culmination was "man".[11] With a focus on fossilized reptiles, Broom was also involved in palaeoanthropology, a burgeoning field in South Africa due to the discovery of hominin fossils that would allow figures such as Raymond Dart and Jan Smuts to claim a pre-eminent position for the country in human origin and evolution. The Boskop Skull had been unearthed in the Transvaal in 1913 and Broom was one of the scientists who examined the skull, determining that "it [had] an enormous brain and is not at all ape-like".[12] For Broom, despite the significance of the find, it was hardly the superman but instead "a Proto-Bushman", an ancestor of the San, a race also believed to be doomed ultimately to extinction.[13]

Broom's view of the San as "living fossils" was akin to that of his contemporary, Raymond Dart, who headed the Department of Anatomy at the University of the Witwatersrand, a department that during the interwar period, included Jack Penn as a part-time lecturer.[14] Introduced to face masks in the 1930s, Dart became convinced of their value in demonstrating physical racial differences as evident in facial features. In 1936, he would initiate the practice in South African physical anthropology of making face masks of various racial groups, in this instance, a San group in the Kalahari. Subjects had reeds placed in their nostrils to allow them to breathe as their faces were encased in plaster of Paris. Their casts would form part of more than a thousand marks that are housed in the Raymond Dart Collection of African Life and Death Masks. For Dart, facial features—physical evidence of racial difference—reinforced his desire to determine "pure" racial types.[15]

While lecturing at Wits, Penn would also engage in studying facial distinctions based on race, addressing the "musculature of the Bantu face" in relation to its European counterpart. An analysis of 135 faces led to finding differences in relation to blood supply, the "cartilages of the

[11] "Stature May Halt, Brain Will Decrease".

[12] Robert Broom, "The Boskop Skull" (Letter to the Editor) in *Nature*, No, 2929, Vol. 116, 19 December 1925, p897.

[13] Broom, "The Boskop Skull", p897.

[14] Jack Penn, *The Right to Look Human: An Autobiography* (Johannesburg, Hugh Keartland Publishers, 1976), p64.

[15] Christa, Kuljian, *Darwin's Hunch: Science, Race and the Search for Human Origins*, (Johannesburg, Jacana Media, 2016) pp61–67.

nose" and the Risorius muscle responsible for facial expression.[16] Facial features—and ultimately the reconstructive surgery to restore them—thus went to the heart of racial thinking in pre-War South Africa. This was buttressed by the origins of plastic surgery itself during the First World War and on the Western Front. Both occurred within a tradition of western medicine that affirmed the white male body as the norm.

This chapter explores the relationship between race and reconstructive surgery in South Africa during the period of segregation and war. It uses the Brenthurst Military Hospital and the 110 Military Hospital at Voortrekkerhoogte to address the hierarchy of expertise between South Africa and Britain, the allocation of medical resources and the contradictions between racial ideology and the humanism of western medicine. The chapter draws largely upon correspondence and reports housed at the Department of Defence Archive and the writings of plastic and reconstructive surgeon Jack Penn.

There was a narrow conception of who constituted the South African soldier during the Second World War and the Union Defence Force (UDF) mirrored the racial and gendered hierarchies of South African society. The key participants in the War were white men with white women and black men relegated to auxiliary roles. White men were envisaged as the ideal combatants with the right to bear arms in defence of the Union. White women were subject to pervasive views of gendered roles that excluded them from combat. Black men were disenfranchised, and opposition existed to arming them, based on an historical relationship of violence, conquest and dispossession. The identity of South African regiments was drawn in large part from the subjugation of indigenous groups and white masculinity was thus defined in relation to its "other", black men.[17] In contrast, even so-called martial races such as the Zulu—who were believed to have a natural inclination to military service and discipline—were armed with assegais rendering them largely impotent in modern warfare.[18] While the exigencies of the frontlines may have blurred

[16] Penn, *The Right to Look Human*, p64.

[17] Suryakanthie Chetty, "'A White Man's War': Settler Masculinity in the Union Defence Force, 1939–1945", *Africa and World War II*, J.A. Byfield, C.A. Brown, T. Parsons and A.A. Sikainga, eds (New York, Cambridge University Press, 2015) pp314–315.

[18] Louis Grundlingh, "The Military, Race and Resistance: The Conundrums of Recruiting Black South African Men during the Second World War", *Africa and World*

the lines between combatant and auxiliary, segregation was vehemently maintained on the home front and this was just as evident in the provision of medical services. Simultaneously, medical training in the relatively new field of plastic and reconstructive surgery was—as had been the case during the First World War—centred on Britain.

Jack Penn was born in Cape Town in 1909 and studied at the University of the Witwatersrand Medical School where he was exposed to the teaching of Raymond Dart. He completed his internship abroad—travelling to both the United Kingdom and the United States—before he returned to South Africa in 1938. According to his autobiography, he was convinced that war would break out in Europe, and South Africa would inevitably be drawn into the conflict. He thus joined the UDF as a part-time officer, holding the rank of major in the 7th Field Ambulance and subsequently the No 2 Medical Officers Training Ambulance of the South African Medical Corps (SAMC).[19] Less is known of Norman Petersen. Born in August 1896, Petersen completed his undergraduate degree at the University of London in 1920. He enlisted in the SAMC in late May 1940 with the rank of Captain before being promoted to Acting Major in November 1940 just prior to his posting to England.[20]

In September 1940, the Director General of Medical Services (DGMS), Edward Thornton, considered the feasibility of having two medical representatives sent to England for training in reconstructive and plastic surgery.[21] Thornton had some knowledge of the field. During the First World War, he had been stationed at the South African Military Hospital in Richmond, England, which focused on the rehabilitation of

War II, J.A. Byfield, C.A. Brown, T. Parsons and A.A. Sikainga, eds (New York, Cambridge University Press, 2015), pp81–84.

19 Jack Penn Service File, W.R. 12/6640, DOD.

20 Norman Petersen Personnel File Reg No 16181, DOD and "Petersen, Norman August Marais", *Plarr's Lives of the Fellows*, Royal College of Surgeons of England, https://livesonline.rcseng.ac.uk/client/en_GB/lives/search/detailnonmodal?qu=LIVES_OCCUPATION%3D%22Maxillofacial+surgeon%22&qf=LIVES_HONOURS%09Titles%2FQualifications%09MRCS+1920%09MRCS+1920&d=ent%3A%2F%2FSD_ASSET%2F0%2FSD_ASSET%3A378202~~0&ic=true, accessed 6 May 2021.

21 Maj-Gen C.G.S. to D.G.M.S., 27 September 1940, File 129/376: Plastic Surgery: MOs Studying in England, Director General Medical Services, Box 92, DOD.

wounded soldiers and in particular, amputees.[22] In addition, Thornton had also been associated with Sidcup making him more than conversant with the value of reconstructive surgery in treating war injuries.[23] Norman Petersen was the choice for one of the two surgeons.[24] He was despatched to England in November 1940 and would return to South Africa in June the following year, whereupon he was assigned to Voortrekkerhoogte.[25]

Penn was the second candidate and a year after his return to South Africa, he once again found himself on a ship to Britain, this time with the aim of gaining experience at the various hospitals established for reconstructive surgery.[26] He would work briefly at Basingstoke, East Grinstead, St Albans and Alton and Stoke Mandeville.[27] Park Prewett Hospital at Basingstoke was under Gillies and contained surgical units for various members of the Allied countries such as the American and Canadian units. It dealt mainly with injured Navy personnel. Fellow New Zealander, Arthur Rainsford Mowlem, ran Hill End Hospital at St Albans, focused on Army personnel as well as civilian casualties. Hill End was also involved in trials testing the efficacy of penicillin in treating infection.[28] The only one of the "Big Four" plastic surgeons to have worked with Gillies at Sidcup and the only one native to the United Kingdom was Thomas Pomfret Kilner who ran Stoke Mandeville that had a high proportion of civilian casualties and was considered by Penn to be the "backwater of plastic surgery at that time".[29] The other three surgeons were native to New Zealand and thus emblematic of the collaboration of the Dominions during the War. In Penn's assessment of the "Big Four", it was clear

[22] "Follow the Drum: South African Military Hospital", *The Library Blog*, 8 November 2017, https://libraryblog.lbrut.org.uk/2017/11/south-african-military-hospital/, accessed 25 November 2020.

[23] "Death of Sir Edward Thornton; Long Service in Public Health", *Rand Daily Mail*, 28 October 1946.

[24] Major, DGMS, 29 October 1940: Captain N. Petersen, SAMC, File 129/376: Plastic Surgery: MOs Studying in England, Director General Medical Services, Box 92, DOD.

[25] Norman Petersen Personnel File Reg No 16181, DOD.

[26] Penn, *The Right to Look Human*, p65.

[27] Penn, *The Right to Look Human*, pp66–67.

[28] M.C. Meikle, "The Evolution of Plastic and Maxillofacial Surgery in the Twentieth Century: The Dunedin Connection", *The Surgeon*, 4, 5, 325–334, p333.

[29] Penn, *The Right to Look Human*, p71.

that he admired Gillies but reserved his real affection for McIndoe (who would give a lecture at the University of the Witwatersrand in 1947) and Kilner who would subsequently be made an honorary member of the Association of Plastic Surgeons of Southern Africa.[30]

Their training completed, both Penn and Petersen returned to South Africa to put their reconstructive surgical skills to use on members of the UDF—and those from other Allied countries. Petersen was stationed at Voortrekkerhoogte for a year before being transferred to Springfield Hospital in Durban and then the Egypt Military Hospital in Cairo until his discharge in 1944.[31] Penn would be stationed at Brenthurst, which became the centre for plastic and reconstructive surgery in South Africa.

In 1941, mining magnate Ernest Oppenheimer donated his home "Brenthurst" in Johannesburg, to the war effort as a Red Cross hospital.[32] Amidst 45 acres of parkland in Johannesburg's exclusive Parktown, the house had been designed by architect Herbert Baker—also responsible for the Union Buildings in Pretoria.[33] Extensive remodelling was undertaken in order to convert the home to a hospital with space for up to fifty patients—many of whom began arriving a mere two months later and were drawn from both South Africa and the other Dominions.[34] Reading about Oppenheimer's donation in a newspaper upon his return to South Africa, Penn requested the use of Brenthurst as a plastic surgery unit from the DGMS. Final approval was given by the Oppenheimers and Penn became Officer-in-Charge of the Brenthurst Red Cross Military Hospital for Plastic Surgery.[35] Eventually another five private homes would be used for reconstructive surgery with Brenthurst remaining the focal point where surgeries were carried out.[36] Brenthurst and its successor Tara admitted only white patients. Those categorized

[30] Penn, *The Right to Look Human*, pp67, 70, 71; Advertisement, *Rand Daily Mail*, 26 July 1947 and "Obituary Notices: T Pomfret Kilner", *British Medical Journal*, 2, 127, 11 July 1964, https://www.bmj.com/content/2/5401/127, accessed 10 December 2020.

[31] Norman Petersen Personnel File Reg No 16181, DOD.

[32] "Vanity Fair", *Rand Daily Mail*, 1 May 1941, p5.

[33] Anthony Hocking, *Oppenheimer and Son* (Johannesburg, McGraw-Hill Book Company, 1973), pp86, 106.

[34] "Vanity Fair", 1 May 1941 and "Vanity Fair", *Rand Daily Mail*, 5 July 1941, p5.

[35] Penn, *The Right to Look Human*, pp73–74.

[36] Hocking, *Oppenheimer and Son*, p195.

as "non-European" with facial injuries were sent to No 110 Military Hospital Voortrekkerhoogte.[37]

Voortrekkerhoogte was a study in contrasts. It had its origins in the British occupation of Pretoria in 1900 during the South African War and was known initially as Roberts Heights, named after Lord Roberts, the commander of the British forces. With the arrival of large numbers of British soldiers, a decision was eventually taken the following year to establish a "Military Cantonment" that would make provision for all the requirements for the army from stores to medical services. The various buildings comprising the early camp were rudimentary and temporary. By 1902 however, wood and corrugated iron buildings were imported from Britain and India and erected on site. This included the first hospital. The cantonment fell under the Union Government during the First World War and served as a prisoner-of-war camp as well as a training site for soldiers of the UDF.[38] For Afrikaner Nationalists, Roberts Heights presented a very "British atmosphere", described as a "caricature of a miniature Aldershot". During the centennial commemoration of the Great Trek—a founding moment in Nationalist mythology—it was subsequently renamed Voortrekkerhoogte.[39] During the Second World War, the hospital based here became a site drawing upon Dominion medical expertise situated within a segregationist South African context.

The creation of the unit for maxillo-facial injury for black soldiers at 110 Military Hospital at Voortrekkerhoogte in 1942 can be contextualized by a changing attitude towards medical care under Jan Smuts's leadership and evident in the Gluckman Commission. The National Health Commission under Henry Gluckman was established in 1942 to investigate the feasibility of creating a national health care system that would include all races within its ambit. During the two years of its existence, the Commission travelled the country, conducting interviews with hundreds of South Africans from medical professionals to representatives

[37] A.J. Orenstein, DGMS: D.G.M.S., U.D.F. Medical Administrative Instruction, No. 137, 9 February 1945, File MD 580/2: Tara: Red Cross Hospital for Plastic Surgery—General, Surgeon General, Group 2, Box 1136, DOD.

[38] Paul J. Els, *Chronicles of the Heights, Vol 1* (Valhalla, Pelsa Books, 2012), pp12–18, 22–23, 30.

[39] Ian Van Der Waag, *A Military History of Modern South Africa* (Johannesburg and Cape Town, Jonathan Ball Publishers, 2015), pp164–165.

of the various racial groups. In its report issued in 1944, the Commission's vision of reform was breath-taking with the recommendation of a universal and non-racial health care system that would be funded by a "common health tax". It was its very ambitious nature that led to Smuts's government ultimately rejecting many of the measures proposed. It is, however, not the recommendations of the Commission but its motivation that have been the subject of historical debate. For Bill Freund, the Gluckman Commission was part of existing efforts under Smuts to foster a "developmental state"—reforms in health were concomitant with those in education and social welfare and necessary for the creation of a modern state, even if they spelled some discomfort for proponents of segregation, Smuts included.[40] According to Shula Marks however, the Commission was a necessary response to the crisis in health care precipitated by the Great Depression and accompanying drought in the previous decade that had had adverse effects on the health of both blacks and whites. The existing, poorly funded and provincially-based health care system proved wholly inadequate to the task and it was largely at the behest of medical professionals, particularly those involved in social medicine, that led to the formation of the National Health Commission.[41] The establishment of the Maxillo-Facial Unit therefore lies at the convergence of the idealism, ideology and pragmatism of medical care in wartime South Africa.

From the outset during the Second World War, a key logistical problem at Voortrekkerhoogte was the maintenance of segregation. Lt. Colonel Lewis Stevens Robertson was forty-seven years old when he enlisted in 1940 and was subsequently named the Officer-in-Command of the Military Hospital, Voortrekkerhoogte where he would serve for the duration.[42] Soon after his posting he reported that the great majority of

[40] Bill Freund, "The South African Developmental State and the First Attempt to Create a National Health System: Another Look at the Gluckman Commission of 1942–1944", *South African Historical Journal*, Vol 64, No 22, 170–186, 2012, pp170–174.

[41] Shula Marks, "Reflections on the National Health Services Commission: A Response to Bill Freund and Anne Digby on the Gluckman Commission", *South African Historical Journal*, Vol 66, No 1, 169–187, 2014, pp174–177.

[42] Lewis Stevens Robertson Service File, W.R. 12/1284, DOD.

"non-European"[43] civilian patients were incurring "considerable unpleasantness" when they were "treated in the same room as Europeans".[44] The situation was little improved by February the following year when Robertson drew attention to the insufficient quality of medical care for "non-Europeans" who were given little in the way of basic facilities, "Four marquees have been erected in the hospital grounds for accommodation of non-Europeans requiring hospitalisation; considerable difficulties are encountered in the running of the non-European sick tents, as there are no side-rooms or proper lavatory or sluice room facilities".[45]

Compounding the lack of facilities—especially for black patients—was the staff shortage where Robertson observed at the outset "that the vast majority of the Hospital staff is almost totally untrained". Military policy also involved what to him seemed the haphazard transfer of existing staff without consultation and with little consideration of medical requirements.[46] Staffing shortages exacerbated the segregation imperative with his protest that, until sufficient black orderlies could be trained, he was compelled to continue using "probationer [white] nurses…in the non-European wards"—including the 110 Military Hospital, established to treat maxillo-facial injuries sustained by black military personnel.[47]

[43] A brief note on nomenclature: "Non-European" referred to those classified as African, Indian or coloured while "European" referred to those classified as white, whether South African or of other nationalities. Unless using the categorization as evident in the primary sources, I utilize the descriptors, "white" and "black".

[44] Letter by Lewis S. Robertson, Lt Col, SAMC, The Officer Commanding, Military Hospital, Voortrekkerhoogte to D.G.M.S. "Medical Attention: Non-Europeans", 22 November 1940, File MD 14/1/9: "Military Hospital V.T.H. arrangements re admission of patients for operative and medical treatment", Surgeon General, Group 2, Box 109, DOD.

[45] Letter from Lewis S. Robertson, Lt Col, SAMC, The Officer Commanding, Military Hospital, Voortrekkerhoogte to the Assistant Director of Medical Services, Voortrekkerhoogte and TVL Command, 24 February 1941, File MD 14/1/9: "Military Hospital V.T.H. arrangements re admission of patients for operative and medical treatment", Surgeon General, Group 2, Box 109, DOD.

[46] Letter from Lewis S. Robertson, Lt Col, SAMC, The Officer Commanding, Military Hospital, Voortrekkerhoogte to D.G.M.S., 27 March 1941, File MD 14/1/2 (Vol 2): "Military Hospital Voortrekkerhoogte Establishment", Surgeon General, Group 2, Box 115, DOD.

[47] Letters from Lewis S. Robertson, Lt Col, SAMC, The Officer Commanding, Military Hospital, Voortrekkerhoogte to D.G.M.S., 1 October 1943 and 20 October 1943, File MD 14/1/2 (Vol 2): "Military Hospital Voortrekkerhoogte Establishment", Surgeon General, Group 2, Box 115, DOD.

Robertson's dilemma can be understood by the prevailing attitude vis-à-vis white women nursing black men that took on a particular valency in the 1930s. Drought and depression had increased the urbanization of both rural Afrikaners and Africans, leading to a resurgence in a "moral panic" regarding the interaction between white women and black men. It is in this light that conservative Afrikaner organizations were vociferous in their opposition to white nurses administering to male black patients with the view that black "servants" should be so employed. This view was not without its detractors, especially among the medical profession, with criticism aimed "at a mentality apparently unable to regard non-Europeans in human terms..." Yet, as Shula Marks argues, the argument was less about race than it was about the use of untrained servants rather than orderlies.[48] Robertson's solution then was to retain the use of nurses until black orderlies were available.

With the establishment of the Brenthurst Military Hospital and 110 Military Hospital at Voortrekkerhoogte, a month-long inspection of South African military hospitals was carried out by Major A.H.R. Champion, a maxillo-facial surgeon in the Royal Army Medical Corps (RAMC) in September 1942. Champion's first stop was Brenthurst—where he would spend the bulk of his time. Brenthurst was likely of particular interest to Champion due to its treatment of imperial troops, who formed three quarters of the patients at the hospital and "were found to be very cheerful and well looked after...[with] ample facilities for occupational therapy". Patients were housed in small wards—of which there were six. There was also the recommendation of an additional "Dental hut" to be constructed on the grounds with a capacity of up to forty beds—that would be funded by Ernest Oppenheimer—to handle the increasing number of cases envisaged from the Middle Eastern theatre of war. Also agreed to would be the employment of additional dental surgeons.[49]

Another aspect of Champion's assessment of Brenthurst was its technological innovation where "acrylic moulds" were used for "grafting eye sockets". Mentioned too was the Brenthurst Splint that promised to be an improvement to the Roger-Anderson Pin currently being used. The

[48] Shula Marks, *Divided Sisterhood: Race, Class and Gender in the South African Nursing Profession* (Johannesburg, Witwatersrand University Press, 1994), pp60–61.

[49] A.H.R. Champion, "Report on Tour of Union of South Africa", File MD 580 (Vol 2): Treatment of Facial Injuries Maxillo-Facial Treatment, R.A.M.C, Group 2, Box 1101, DOD.

Brenthurst Splint had been funded to the amount of £12,500 by the Union Government.[50] A meeting of medical and military personnel was also held at Brenthurst during Champion's visit where Penn showed a film depicting surgical work carried out at Brenthurst. In light of the free exchange of knowledge, Champion also presented on "early maxillofacial wounds in patients" and ultimately deemed the gathering of like minds "very successful".[51] As such—and as discussed in the following chapter—both world wars created the opportunities for innovation and invention, especially in a specialization that was relatively new. It was therefore Penn's work on the injured soldiers from the East and North African theatres of war that allowed for experimentation: the development of new techniques and technology.

In contrast to the almost two weeks at Brenthurst (with various other inspections interspersed), Champion would only spend the 5th of September at Voortrekkerhoogte, named as the "Robert Heights Military Hospital Dental Centre" in his itinerary. His observations were brief where he described the training and experience of staff, including Lester Brown and Harry Goldin, considered the hospital to be "well equipped with excellent lighting" and mentioned the current number of patients being treated: "6 European and 9 non-European". Champion also visited another hospital addressing the treatment of black patients, the Springfield Military Hospital in Natal. Treatment was provided by the resident reconstructive surgeon, Lt. Col. Norman Petersen whose ambit included surgical work on the face, hands as well as jaws. Champion thus considered his visit a success, recommending "the advantages of climate and surroundings" for recuperating soldiers. He also suggested the "centralization of all fractured jaws and plastic cases" for prompt and effective treatment.[52]

In conjunction with Champion's visit and anticipating a possible increase in the number of patients sent to South Africa from the Middle East, a proposal was put forward for the expansion of Brenthurst and use of an auxiliary hospital that would increase the number of beds, supplement the necessary equipment and make available the services of

[50] The invention and utilization of the Brenthurst Splint is discussed in greater detail in the subsequent chapter.

[51] Champion, "Report on Tour of Union of South Africa".

[52] Champion, "Report on Tour of Union of South Africa".

a dental officer at short notice. While black patients were still to be allocated to Springfield or Voortrekkerhoogte, in the case of Brenthurst being overwhelmed, injured white soldiers would be treated at Springfield "with the consequent necessity for provision for non-Europeans at Voortrekkerhoogte".[53]

The exigencies of war notwithstanding along with staffing and space shortages, there was an early perception of Voortrekkerhoogte as a site of training, if not innovation. Particularly in relation to plastic surgery, the military hospital was very much the junior partner in its relationship with Brenthurst, dealing successfully with minor surgery while calling on Brenthurst expertise for serious cases. Writing to the DGMS, Lewis Robertson requested that in addition to junior officers, his senior medical staff also be given the opportunity for training in pioneering medical techniques which could then be applied at the medical hospital in Voortrekkerhoogte. One of these was Captain C.A.R. Schulenburg with the request made for him to be allocated time at Brenthurst so that he could acquire skills that could be used to "relieve" the strain on Brenthurst but also mean that Voortrekkerhoogte could "perfect" its plastic surgery work.[54] Carl August Richard Schulenburg had enlisted in his mid-twenties and would spend the great part of his war service at Voortrekkerhoogte where he was a surgeon addressing "septic traumatic surgery; burns, lacerations [and] septic compound fractures". This made him an ideal candidate for the acquisition of more complex surgery skills, and he was assigned briefly to Brenthurst in the wake of Robertson's request.[55] The skilling of Schulenburg was an early step in the process of making Voortrekkerhoogte a centre for reconstructive and plastic surgery.

Another step was the maintenance of ties with the home of modern plastic and reconstructive surgery. As 110 Military Hospital was establishing itself as a centre for maxillo-facial surgery, the hierarchical ties

[53] Director of Dental Services, Medical Directorate, Pretoria, "Provisions Made for Special Treatment", 11 September 1942, File MD 580 (Vol 2): Treatment of Facial Injuries Maxillo Facial Treatment, Surgeon General, Group 2, Box 1101, DOD.

[54] Letter by Lewis S. Robertson, Colonel, Commanding Military Hospital: Voortrekkerhoogte to DGMS, "Exchange of Officers", File MD 14/1/2 (Vol 2): Military Hospital Voortrekkerhoogte Establishment, Surgeon General, Group 2, Box 110, DOD.

[55] C.A.R. Schulenburg Service File, W.R. 12/11663 and Letter from Lewis S. Robertson to Assistant Director Medical Services, File MD 14/1/2 (Vol 2): Military Hospital Voortrekkerhoogte Establishment, Surgeon General, Group 2, Box 110, DOD.

with British expertise and South African surgery remained. Captain Lester Brown enlisted in the UDF at the relatively advanced age of 38 with an early posting as a lieutenant in the South African Medical Corps (SAMC) training centre in Sonderwater. As a dental officer, he was attached to the military hospital at Voortrekkerhoogte but also served as the "maxilla facial specialist" at Brenthurst.[56] In 1943, Brown reminded the DGMS of the sending of two plastic surgeons and two dental surgeons to the reconstructive surgical centres in England in early 1940 by the various Dominions. He lamented South Africa's failure which "for some unknown reason was the only Dominion which did not include dental surgeons in its team" and further did not rectify the exclusion even after a request was made by Penn and Petersen upon their arrival in England. Brown highlighted the relevance and importance of studies conducted at Rooksdown House, East Grinstead and St Albans in addressing infection and the immobilization of the jaw that had taken on a greater relevance given the "new and important demands" on dental surgery due to wartime injury. As such, Brown requested the placement of two dental surgeons for four to six months on a course that would familiarize them with the methods being developed and tested in England. The aim was for them to acquire the "highest degree" of skill in the treatment of the injured. While acknowledging the contributions of Brenthurst—including the non-invasive Brenthurst splint designed by Penn and Brown to immobilize the jaw, allowing healing and reducing infection—the work of the British centres of plastic surgery was portrayed as invaluable in contributing to South African expertise, "We feel that we have much to learn in this specialised branch, from our overseas colleagues who have had a vast amount of experience not only in the last war, but in this one too".[57]

[56] "Brown, Leslie Julius", Service Card No W.R.12/7091, 176921, DOD. There is some discrepancy in terms of Brown's first name with various publications emanating from Brenthurst accredited to "Lester Brown". His service card however, held at the Department of Defence Archives, indicates that his name is "Leslie Brown".

[57] Letter from L.J. Brown, Captain SAMC, to DGMS, 7 May 1943, File MD 580 (Vol 2): "Treatment of Facial Injuries Maxillo-Facial Treatment", Surgeon General, Group 2, Box 1101, DOD. For more detail on the development of the Brenthurst Splint please refer to Jack Penn and Lester Brown, "A New Conception in Extra-Oral Splintage", *Brenthurst Papers*, Jack Penn, (ed) (Johannesburg, Witwatersrand University Press, 1944), Brn. No: 3297, Call No: 617.95 BRE, Brenthurst Library.

The process of development was steady, and a newspaper article reported positively on medical developments at Voortrekkerhoogte, describing the ambitious provision of various medical services—ranging from the plastic and reconstructive surgery currently occurring at Brenthurst to other forms of surgery as well as gastric and pulmonary treatment—in one complex. The article, appearing in the *Sunday Times* in January 1943, recognized that Brenthurst—and the other homes donated—would only be available for the duration of the conflict and Voortrekkerhoogte was therefore the site where state of the art medical facilities would be available on a permanent basis. Also highlighted were the activities available for patients during the period of rehabilitation and convalescence including a gymnasium and pool for physical therapy as well as "adult education and occupational and vocational therapy" that bore the hallmarks of DGMS Edward Thornton's influence. The comment on the article by the office of the DGMS however, considered the article to be premature in its assessment of Voortrekkerhoogte and thus "somewhat unfortunately worded".[58]

While Voortekkerhoogte continued to experience various logistical difficulties, by April of that year, the 110 Military Hospital had become the main treatment centre for black military personnel with maxillo-facial injuries. Injuries, treatment and prognosis were recorded on standardized forms. The form (Fig. 5.1) is that of a private in the Native Military Corps (NMC). Along with detailed notes, images of the skull, jaw and the face (right, frontal and left) were used to indicate the site of injury (Fig. 5.1).

Fourteen months after the establishment of the Maxillo-Facial unit in September 1942, Major Harry Goldin could report that the hospital had treated 120 patients with jaw fractures, the great majority of which were fractures to the lower jaw. Most injuries were a result of fighting between soldiers and due to the "army boot". According to Goldin, "The non-European goes completely berserk when roused to fight [and] loses all control", thus resulting in severe and multiple facial injuries. These injuries were treated through open surgery and the use of wiring as there had been a shortage of "extra oral splints" conventionally used to mobilize the jaw. In addition to jaw fractures, Goldin also highlighted the unit's work in addressing other kinds of injuries related to the jaw and

[58] Letter (with article attached) from DGMS to the ADMS, Northern Command, "Special Hospital for Soldiers", 26 January 1943, File MD 14/1 (Vol 1): Military Hospital Voortrekkerhoogte General, Surgeon General, Group 2, Box 111, DOD.

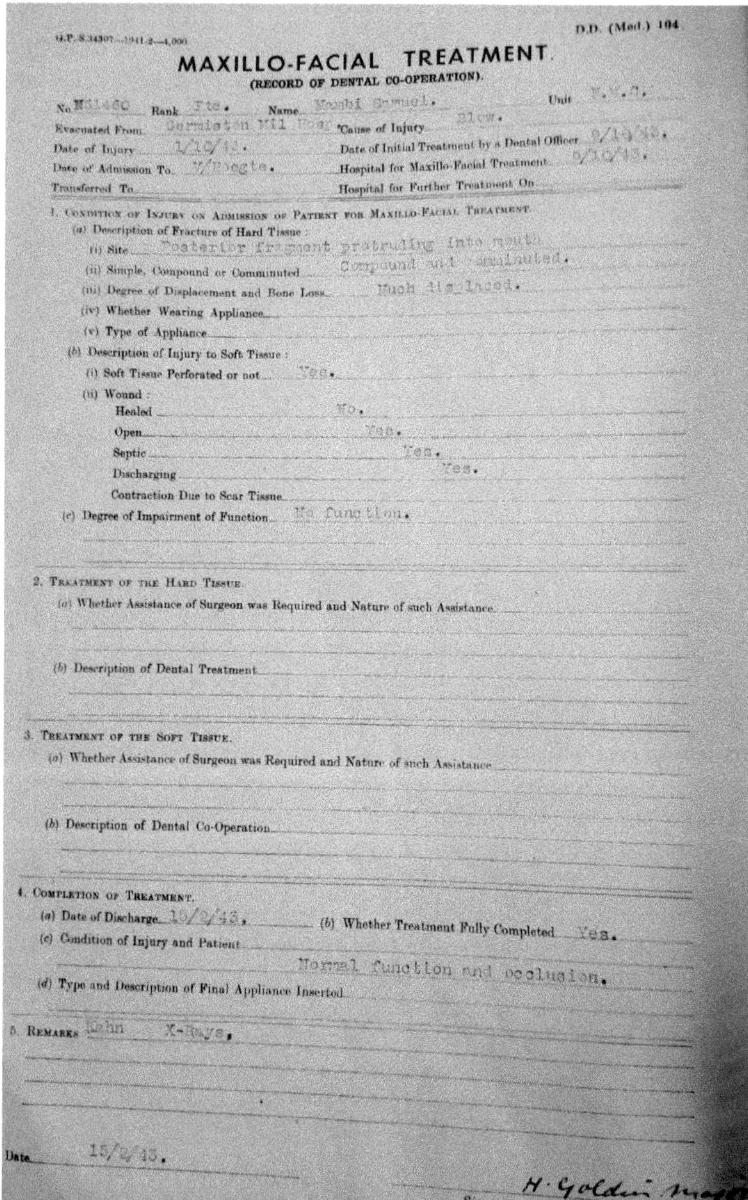

G.P.-S.34307—1941-2—4,000

D.D. (Med.) 104

MAXILLO-FACIAL TREATMENT.

(RECORD OF DENTAL CO-OPERATION).

No. N51460 Rank Pte. Name [illegible] Samuel. Unit [illegible]

Evacuated From Germiston Mil Hosp. Cause of Injury Blow.

Date of Injury 1/10/4[illegible]. Date of Initial Treatment by a Dental Officer 9/10/43.

Date of Admission To V/[illegible]. Hospital for Maxillo-Facial Treatment 9/10/43.

Transferred To Hospital for Further Treatment On

1. Condition of Injury on Admission of Patient for Maxillo-Facial Treatment.

(a) Description of Fracture of Hard Tissue:

(i) Site Posterior fragment protruding into mouth

(ii) Simple, Compound or Comminuted Compound and comminuted.

(iii) Degree of Displacement and Bone Loss Much displaced.

(iv) Whether Wearing Appliance

(v) Type of Appliance

(b) Description of Injury to Soft Tissue:

(i) Soft Tissue Perforated or not Yes.

(ii) Wound:

Healed No.

Open Yes.

Septic Yes.

Discharging Yes.

Contraction Due to Scar Tissue

(c) Degree of Impairment of Function No function.

2. Treatment of the Hard Tissue.

(a) Whether Assistance of Surgeon was Required and Nature of such Assistance

(b) Description of Dental Treatment

3. Treatment of the Soft Tissue.

(a) Whether Assistance of Surgeon was Required and Nature of such Assistance

(b) Description of Dental Co-Operation

4. Completion of Treatment.

(a) Date of Discharge 15/2/43. (b) Whether Treatment Fully Completed Yes.

(c) Condition of Injury and Patient Normal function and occlusion.

(d) Type and Description of Final Appliance Inserted

5. Remarks [illegible] X-Rays.

Date 15/2/43.

H. Goldin Major

Signature of Dental Officer

Fig. 5.1 Surgeon General, Box 1135, Group 2, MD 580: Treatment of Facial Injuries, Case file 15 February 1943, DOD

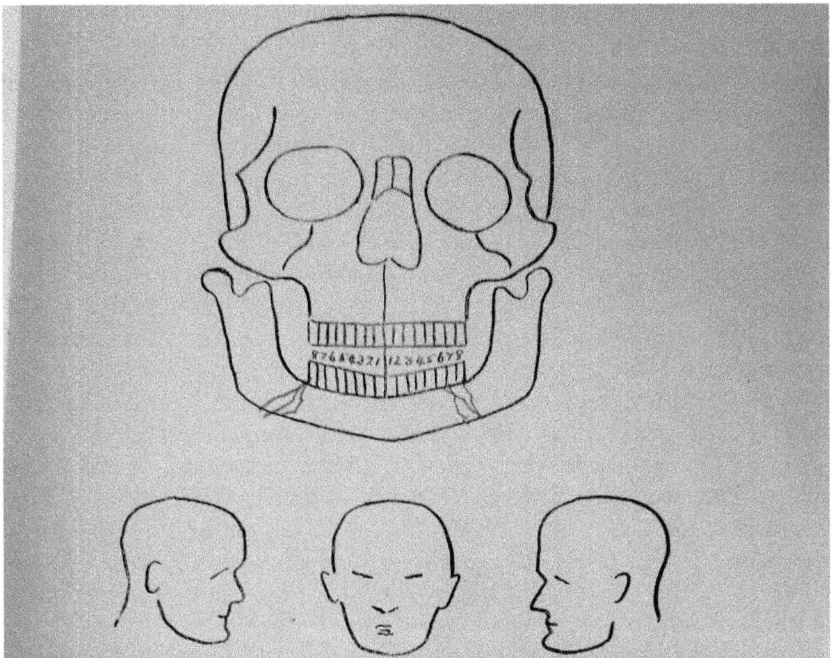

SKETCH OF APPLIANCE :

Exam. Great pain. Severe swelling at angle 8 7 raised with posterior fragment. very septic. Pouring puss from fracture site 8/ 8 . Much Buccal displacement. of left angle fracture and raised posterior fragment. 11/10/43 impressions. Jaws wired together. 12/10/43 Further lower impressions. Jaws wired together. 16/10/43 Ext. 87. Extention 87 . Elastic traction. 21/10/43 Precision locked. 2/11/43 Removed extention No pus. Discharge from mouth. No swelling. Jaws immobilised.

PHOTOGRAPHS OF PATIENT BEFORE AND AFTER TREATMENT (IF POSSIBLE).

Before. **After.**

5/11/43 Smooth rough edge where extention was removed. Re-lock. 20/11/43 Reset lower splint and precision locks. 25/11/43 Examination 26/11/43 Release jaws. Non-Union 8/ 78 . overlap /78 . Pain. To have extra oral splint. 29/11/43 Open operation on mandible. Freshening of Fracture and extra oral splints. Both sides. 10/12/43 Precision locked. 15/12/43 Examination. 28/12/43 Release jaws. 29/12/43 Exam. 31/12/43 Reset fish plates 4/1/44 Release jaws. 8/1/44 Remove hook. 10/1/44 Check extra oral splint. Left back. Clamps very loose. 11/1/44 Exam. Polish splints. 22/1/44 Remove lower splint. 24/1/44 Examination. 25/1/44 Extra oral splints removed under G.O. Anes. Union at left angle not complete. 28/1/44 New lower metal cap splint. Impressions taken and X-Rays. 29/1/44 Exam and gums painted with iodine. 31/1/44 Lower metal splint cemented. 3/2/44 Adjustment 11/2/44 Splints removed. and proph. 12/2/44 Proph. completed. 15/2/44 Discharged.

Fig. 5.1 (continued)

wrote approvingly of the importance of the unit as a central hub in treating injury incurred by UDF personnel. He had seen much evidence among civilians of poorly diagnosed and treated maxillo-facial injuries that had resulted in "deformities and lack of function". Concluding his report, Goldin was careful to emphasize that the treatment accorded to black soldiers at Voortrekkerhoogte was of the same standard as that received by their white counterparts.[59] Yet the patients were distinctive; while Penn would describe both "battle" and "bottle" injuries—those resulting from combat and those due to fights breaking about between inebriated soldiers[60]—for the non-combatant, auxiliary soldier of the NMC, Goldin suggested a lack of self-control and violence, ultimately making them responsible for their injuries. In addition, there was a further complication—the unwillingness of two patients to accept treatment as they believed themselves to have been "bewitched", rendering western surgical methods ineffective. In the conflict between medicine and indigenous belief, these two patients were released from military service due to being found psychologically unfit.[61] Writing in 1977, anthropologist Harriet Ngubane's discussion of Zulu practices demonstrates similarities with that observed by the missionaries among the Xhosa a century earlier—the role of the ancestors, polluting elements and sorcery as the causes of afflictions that affected indigenous Africans.[62] Moreover, the confinement of these ailments to the African population leant it a particular Africanist slant that was in some ways impervious to the influence of western medicine. Goldin's discussion of patients who considered themselves to have been "bewitched" was determined by the surgeon to be a sign of mental illness, yet it also offers a tantalizing glimpse of the reactions of African patients confronted with the authority of white, western medical expertise.

The figure emblematic of this medical expertise who further bestrode the two worlds of Brenthurst and Voortrekkerhoogte was Jack Penn who

[59] H. Goldin, Major, SAMC, Maxillo Facial Department, "Report on Non-European Maxillo Facial Centre", 9 November 1943, File MD 580 (Vol 2): Treatment of Facial Injuries Maxillo Facial Treatment, Surgeon General, Group 2, Box 111, DOD.

[60] Jack Penn, "Impressions of Plastic and Maxillo-Facial Surgery in the Palestine War", *Plastic and Reconstructive Surgery*, 6, 4, October 1950, pp327–333.

[61] Goldin, "Report on Non-European Maxillo Facial Centre".

[62] Refer to Harriet Ngubane, *Body and Mind in Zulu Medicine: An Ethnography of Health and Disease in Nyuswa-Zulu Thought and Practice* (London, Academic Press, 1977).

would be requested to pay a monthly visit to the hospital to consult on injuries from his base at Brenthurst.[63] During his tenure at Brenthurst, Penn recorded some of the more memorable cases in the first South African journal of plastic surgery, the *Brenthurst Papers*. While taking its inspiration from Gillies' *Plastic Surgery of the Face*, it was not as ambitious a project, yet Penn followed Gillies' format with the depiction of injury, a sketch of the corrective procedures and the final outcome. And, as with Gillies' experiences on the Western Front, Penn's case studies depicted the facial deformities sustained by white soldiers and success was determined by the extent to which a "normal" aesthetic was restored. In addition, however, articles appeared by various contributors on reconstructive surgical methods used for the face and hands as well as the development and utilization of the Brenthurst Splint.[64]

Penn's acknowledged expertise meant that he was called upon by the military to advise with regard to potential reconstructive surgery cases. An accomplished artist and sculptor, he drew up a general report for treatment to be given to those with facial injuries, complete with rough sketches. This traced the process of treatment from the field to the hospital. At the outset he stressed the unique nature of facial injuries—conventional methods could not be used yet injuries were also not susceptible to gas gangrene and healing was rapid. The first part of the report dealt with the initial provision of first aid soon after injury: treatment of bleeding and shock and the prevention of suffocation and infection. Broken noses could be set with a "finger and thumb" or the use of forceps as a last resort. A fractured lower jaw needed to be wired into position with the same procedure applied to the upper jaw. Once the patient had been taken to a medical station, there should be further cleaning of the wound, and the injury prepared for eventual reconstructive surgery; plaster of Paris and wiring could then be used to further immobilize and position the jaw with a "stockinette" placed over the head. He also provided instructions for the installation of a tube to "irrigate" the bone, should this be necessary and, in a humorous nod to

[63] Letter from Lewis S. Robertson, Lt Col, SAMC, The Officer Commanding, Military Hospital, Voortrekkerhoogte to D.G.M.S., "Facio-Maxillary Cases in Non-Europeans", 9 April 1943, File MD 14/1/14: Military Hospital—Voortrekkerhoogte: Consulting Physicians, Surgeon General, Group 2, Box 108, DOD.

[64] Refer to *Brenthurst Papers*, Jack Penn, (ed) (Johannesburg, Witwatersrand University Press, 1944), Brn. No: 3297, Call No: 617.95 BRE, Brenthurst Library.

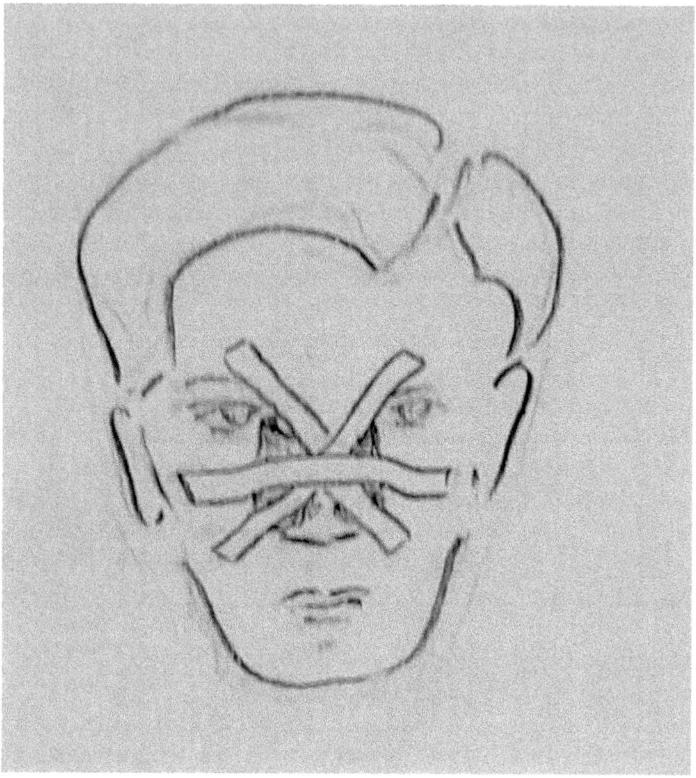

Fig. 5.2 Sketches taken from Jack Penn "Treatment of Facial Injury", Surgeon General, Box 1135, Group 2, MD 580: Treatment of Facial Injuries, DOD

compassion, suggested an opening at the top "to allow the patient to scratch his head". The necessary plastic and reconstructive surgery should only be undertaken at a specialized centre for it. Penn's recommendations were illustrated by rough sketches that used the western medical convention of a white male face[65] (Fig. 5.2).

[65] Jack Penn "Treatment of Facial Injury", File MD 580: Treatment of Facial Injuries, Surgeon General, Group 2, Box 1135, DOD.

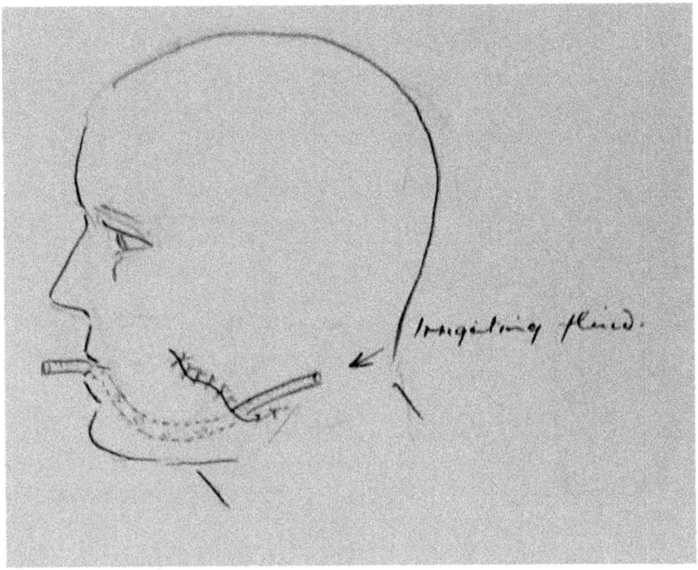

Fig. 5.2 (continued)

Complementing Penn's recommendations was UDF Administrative Instruction 83 issued by Brigadier A.J. Orenstein, Thornton's replacement as DGMS. Under Instruction 83, all white males with jaw fractures and/or requiring plastic surgery were to be sent to Brenthurst and all "non-Europeans" to the Voortrekkerhoogte Military Hospital. Most significant—and the primary instruction—was the following:

> It is of great importance to the ultimate fate of the patient that cases of fracture of the jaw and those requiring plastic operations or extensive skin grafting should be seen and dealt with by Specialist officers as soon as possible after the injury has been inflicted.

Further, if it was not possible due to the nature of the injury to have patients transferred to their allocated hospital, it was essential that the

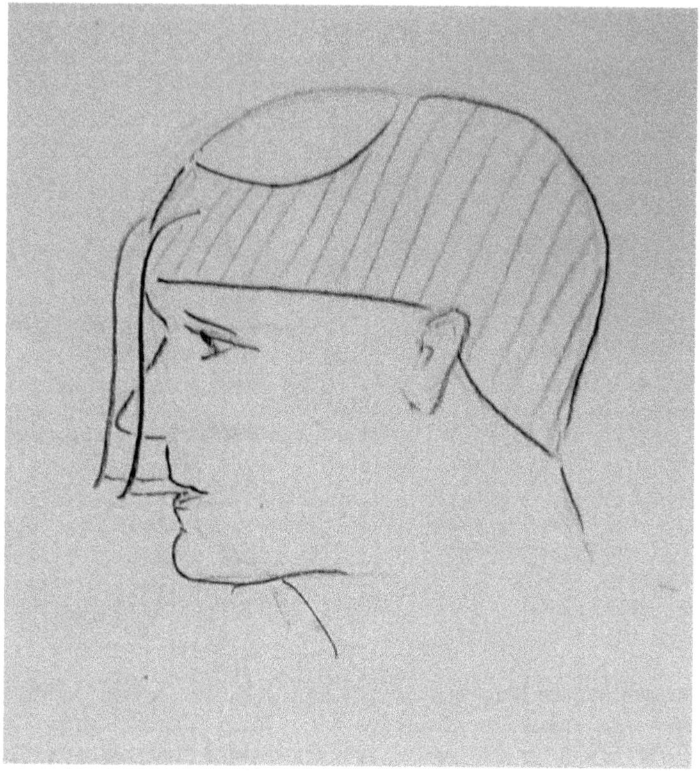

Fig. 5.2 (continued)

Officer-in-Charge at Brenthurst (viz. Jack Penn) be consulted telephonically or via telegram regarding early treatment.[66]

At Voortekkerhoogte a challenge and source of frustration faced by medical personnel—and one that demonstrated that the procedures for effective treatment as proposed by Penn and the DGMS were not necessarily easily implemented in practice—was the delay in the treatment of black patients. Private Elias Gaza enlisted in the UDF in February 1941. Married with five children, he was categorized as a "Shangaan"

[66] A.J. Orenstein, "DGMS UDF Administration Instruction, No 83", 27 August 1943, File MD 580 (Vol 2): Treatment of Facial Injuries Maxillo-Facial Treatment, Surgeon General, Group 2, Box 1101, DOD.

or Tsonga and attested at Tzaneen in present-day Limpopo. He served within the borders of South Africa as a cook with the NMC and would be awarded the Africa Services Medal.[67] In August of 1941, Gaza was admitted to Standerton Hospital as a result of a truck accident. Diagnosed with an injured jaw, he was transferred to the Non-European Hospital in Johannesburg almost three weeks later and subsequently released for duty on the 15th of September. On the 26th of September, complaining of "injury to jaw and severe headache" he was re-admitted to the Non-European Hospital and a week later was returned to duty and treated for sinusitis. Given seven days of sick leave he was subsequently diagnosed with neuritis. Almost two weeks later the diagnosis was amended to a "fractured skull" and "post concussion neurosis". Almost six weeks after that—and four months after the original injury—it was determined that he had a fractured jaw. Jack Penn was consulted on his case on the 13th of January 1942, at which point Gaza unsurprisingly "[refused] treatment and [wanted] to go back to his Unit". In his report, Lester Brown was both clearly outraged at the delay—evident by his detailing of every aspect of misdiagnosis and incorrect treatment—yet circumspect in recommending disciplinary measures against any medical officers to whom blame could be allocated:

> In view of this total disregard of one of the essential services of the SAMC for non-Europeans, I would suggest that with your consideration and recommendation, all non-Europeans requiring Maxillo-Facial treatment be directed to the non-European section of the Dental Hospital...May I suggest, Sir, that this Report and recommendation be treated semi-Officially and will not censure any particular individual in it's [sic] consideration.[68]

More than two and a half years after Brown's complaint, Goldin, attached to the Maxillo-Facial Unit at 110 Military Hospital, cited three cases of individuals injured in Natal where there were delays of anything from a week to almost a month in their transfer to Voortrekkerhoogte. Two of

[67] Elias Gaza Service File, Regt No 13948, DOD.

[68] Lester Brown, Captain, SAMC to Command Dental Officer, Witwatersrand Dental Command, "re: Elias Gaza No: 13948 (Pte) N.M.C.", 22 January 1942, File MD 580: Treatment of Facial Injuries, Surgeon General, Group 2, Box 1135, DOD.

them had had their injuries turn septic as a result.[69] In January 1945, Goldin drew attention to his previous missive, providing another report on two patients experiencing similar delays in treatment. One of these was Private Izak of the NMC who had fractured his jaw in October 1944 and was transferred to 110 Military Hospital "49 days after injury" having had no treatment, despite the early and correct diagnosis of his injury. Upon admittance to Voortrekkerhoogte, Goldin's description of Izak's condition is as follows:

> The left angle of mandible swollen.
> Posterior fragment...much displaced and raised. Patient unable to close mouth.
> Pus pouring from gap at fracture site...
> Complete loss of function.[70]

As 1943 drew to a close, the various measures implemented to transform 110 Military Hospital to a fully functional and permanent unit for maxillo-facial and plastic surgery were beginning to bear fruit. In a report prepared by Lt. Colonel Rose-Innes serving under Lewis Robertson, a summary was given of the steps taken thus far: the early provision of a medical officer familiar with the basic procedures for simple plastic surgery; the allocation of a dental officer to address maxillo-facial injuries; and the drawing upon of expertise from the Brenthurst Hospital and particularly that of Jack Penn who prepared hospital personnel to undertake more ambitious work in reconstructive surgery. Central to this were the "non-European" patients of 110 who were the beneficiaries of the growing surgical expertise. Yet this expertise was largely due to the many prospects afforded by wounded black soldiers and Rose-Innes described the unit as having "one of the finest collections of fractured jaws in the country". These men provided the "splendid 'practice ground' to bring

[69] H. Goldin, Major, SAMC, Maxillo Facial Unit to Medical Officer-in-Charge, Surgical Division, 110 Military Hospital, Voortrekkerhoogte, "re: Transfer of Maxillo-Facial Cases to No 110 Military Hospital, Voortrekkerhoogte", 22 August 1944, File MD 580 (Vol 2): Treatment of Facial Injuries Maxillo-Facial Treatment, Surgeon General, Box 1101, DOD.

[70] H. Goldin, Major, SAMC, Maxillo Facial Unit to Medical Officer-in-Charge, Surgical Division, 110 Military Hospital, Voortrekkerhoogte, "re: Transfer of Maxillo-Facial Cases to No 110 Military Hospital, Voortrekkerhoogte", 3 January 1945, File MD 580 (Vol 2): Treatment of Facial Injuries Maxillo-Facial Treatment, Surgeon General, Box 1101, DOD.

the Maxillo Facial [sic] work up to standard". The result of these measures and the experience obtained from working on black soldiers underlay 110 being able to adequately replace wartime and temporary hospitals, making the former the "chief permanent military hospital in the Union". Foreseeing the likelihood that many of the Dominion soldiers that were being treated at Brenthurst, would now be sent to England as well as the closing down of Brenthurst itself, Rose-Innes proclaimed that 110 Military Hospital was now in a position to treat "European" soldiers as well as their black counterparts.[71] Expertise in treating the former had therefore come from the teething pains associated with treating the latter.

While black patients had to an extent provided the necessary training for plastic and reconstructive surgeons, there was also a particular expertise acquired when treating black soldiers. While Penn's skills could be applied generally to both black and white patients, he also alluded to specific features associated with black patients. Writing almost three decades later, he recorded what he described as "the problem of pigment" in his autobiography. Despite considering briefly the ailments specific to a lack of pigment such as skin cancer, Penn's focus was on the unique challenges presented by pigmented skin for plastic surgeons. The epidermis is associated with skin colour and is only "the width of tissue paper" yet, even when the epidermis is removed, pigmented cells from the hair follicles renew the epidermis and the pigment is particularly prominent in the region surrounding the hair follicle. This darkening of the skin then expands and is noticeably darker than the usual skin tone of the individual. Skin grafting thus presents a challenge as the graft takes on a darker hue and its contrast with the "normal" skin leads to "a patchwork appearance". The use of "full thickness flap transfers" for major reconstructive surgery on black patients also meant that it was difficult to determine in "black or brown skin" if the graft had taken. If circulation was impeded in white skin, the flap went "dead white" but pigmented skin could not "turn dead white" and so limited the surgeon's ability to assess the efficacy of the graft. Through his description of these challenges associated with pigmented skin and his perception of pigment as a "problem", Penn's observations suggest white skin as a norm that was

[71] R.H. Rose-Innes, "Maxillo Facial and Plastic Unit at 110 Military Hospital", 22 October 1943, File MD 580 (Vol 2): "Treatment of Facial Injuries Maxillo-Facial Treatment", Surgeon General, Group 2, Box 1101, DOD.

amenable to the intervention of the surgeon. Concomitantly pigmented skin presented an intractable "problem" a deviation from the norm.[72]

With the War coming to an end, Brenthurst was indeed returned to the Oppenheimers and Penn was tasked by the DGMS with the renovation of a fire-damaged estate that would be converted into a new hospital, Tara. Like Brenthurst, Tara was also run by the Red Cross but under the ultimate control of the UDF. As such, the types of patients permitted for treatment were clearly spelled out—male and female patients that were classified "European" and requiring skin grafts, plastic surgery for disfiguring injuries and those with complex jaw fractures.[73]

In the aftermath of the War, the hospital at Voortrekkerhoogte continued to treat maxillo-facial and other injuries requiring the skills of reconstructive and dental surgeons. The patients were both white and black and fractured jaws remained the common injury. Civilians too were referred from the Pretoria Hospital to the Maxillo Facial unit, now housed at No 1 Military Hospital, Voortrekkerhoogte. The latter were charged a fee, and injuries included the ubiquitous jaw fracture as well the reconstruction of jaws and use of prosthetics due to cancer.[74] Patients' personal details were recorded on a standard form that included sections for remarks made by medical personnel regarding treatment as well as follow-up examinations. Also evident on the form was a sketch of the face and jaw where the injury was graphically illustrated (Fig. 5.3).

Even more so than the diagnostic forms used during the War itself, the new form held true to the convention of the white male face as the standard on which deviations were recorded (Fig. 5.3). It thus rendered invisible the arduous process of training and surgical development undertaken at Voortrekkerhoogte, a development that was based on plastic and reconstructive surgery undertaken on black patients.

From its origins in the First World War, modern plastic and reconstructive surgery was grounded in white male proficiency and collaboration

[72] Penn, *The Right to Look Human*, pp20–21.

[73] A.J. Orenstein, DGMS: D.G.M.S., U.D.F. Medical Administrative Instruction, No. 137, 9 February 1945, File MD 580/2: Tara: Red Cross Hospital for Plastic Surgery—General, Surgeon General, Group 2, Box 1136, DOD.

[74] Command Dental Officer, 1 SA Military Hospital, Voortrekkerhoogte, Report: "Maxillo Facial Department, 1 S.A. Military Hospital", File MD 580 (Vol 2): Treatment of Facial Injuries Maxillo-Facial Treatment, Surgeon General, Group 2, Box 1101, DOD.

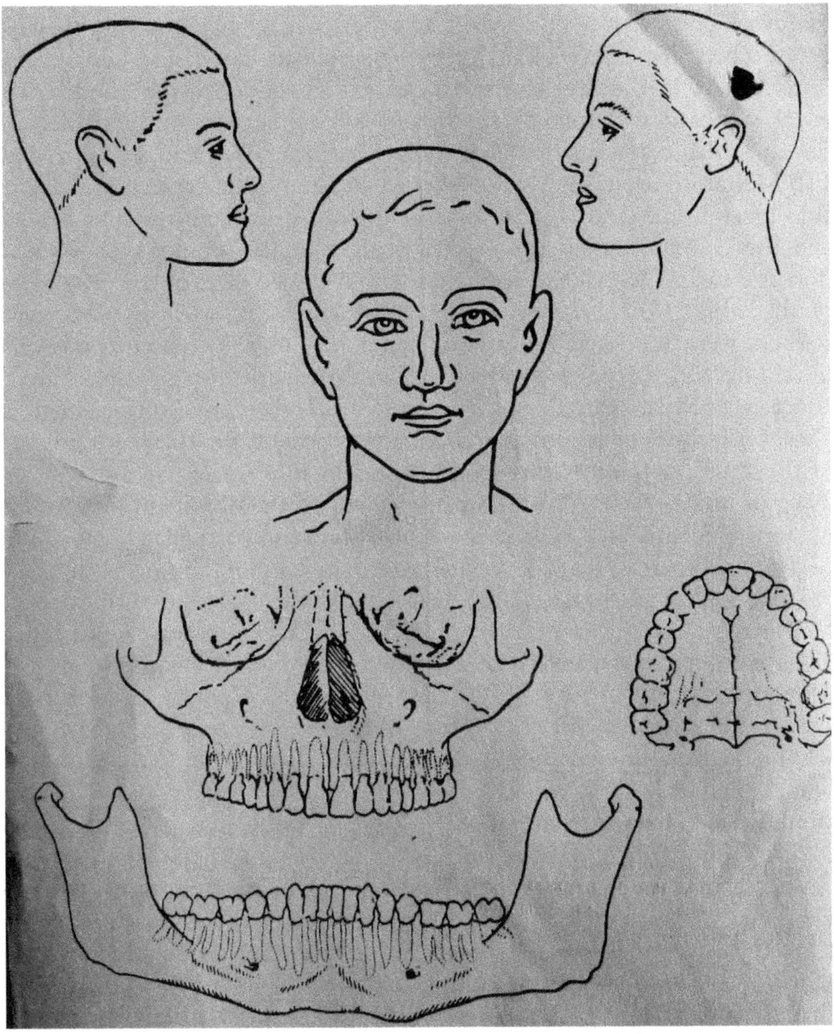

Fig. 5.3 "Maxillo-Facial Treatment", 18 September 1950, Surgeon General, Group 2, Box 1101, MD 580 (Vol 2): Treatment of Facial Injuries Maxillo-Facial Treatment, DOD

that was bolstered by the surgical experience gained on the Western Front where the injuries to white soldiers led to innovation and specialization. Within the South African context in the following conflict, the racial standard was further reinforced by segregation and the disparities in health care and provision of services for black and white troops. 110 Military Hospital at Voortrekkerhoogte demonstrates the uncomfortable tension between the universalist discourse of medicine and healing and the commitment of its medical staff with the ideology of segregation and racial inequality. Occurring within the context of the privileging of the white male body in western medicine, the faces of black male soldiers were rendered a deviation from the norm yet also formed the terrain on which expertise was built. Simultaneously, the genuine sense of concern and real struggles on the part of reconstructive surgical staff at Voortrekkerhoogte to provide adequate treatment for the black soldiers of the UDF, highlights the idealism evident in a society at war where a sense of unity—however tentative—challenged the dominant narrative in western medicine and in reconstructive surgery in particular.

The main part of his war service over, Jack Penn established his private practice, the Brenthurst Clinic, named for the wartime surgical unit. Dominating the foyer of the Clinic was a tapestry created by Ernest Ullmann depicting the almost 5000-year-old history of plastic surgery from ancient Egypt to the contemporary era. Flanked by surgeons, ancient and modern, the focal point of the tapestry was two trees, "one with a broken branch and withered leaves; the other with a grafted branch and flowering leaves". The title of the tapestry—which Penn would use for his autobiography—was "It is the divine right of man to look human".[75] If man had been created in the image of God, plastic surgery had restored his divinity. It was another articulation of Da Vinci's Vitruvian Man and the humanism that underlay it. Yet the conventions of medicine that were evident in the origins and practice of reconstructive surgery and the inequalities—and the challenges to these inequalities—inherent in wartime South African society rendered this image an ambiguous one.

[75] Jack Penn "A Surgeon's Story: It is the Divine Right of Man to Look Human" (unpublished manuscript), Brenthurst Library, Brn No: 12057, 617.95 PEN.

CHAPTER 6

Invention, Compensation and Dissemination at the Brenthurst Red Cross Military Hospital for Plastic Surgery

"Highly commendable" was the assessment of the Inventions Sub-Committee during the weekly meeting held on Wednesday, the 19th of July 1944. The effusive evaluation was applied to two "Brenthurst Inventions": the acrylic eye and the jaw splint.[1] As discussed in the previous chapter, the Brenthurst Red Cross Military Hospital for Plastic Surgery had been established by reconstructive surgeon Jack Penn in a home donated by the Oppenheimers for the duration of the War. Over the course of the War, other private homes were donated for reconstructive surgery and convalescence and additional Union Defence Force (UDF) clinics were established such as that at Voortrekkerhoogte. Brenthurst, however, remained a significant centre, a source of expertise and of innovation. This was due largely to two factors: the attitude and ambitions of its founder, Jack Penn, and the field of reconstructive surgery itself.

As was the case with orthopaedics, reconstructive surgery was birthed as a distinctive specialization during the course of the First World War. Previously, both had been the purview of general surgeons however, the range of projectile-induced injuries resulting from trench warfare impelled their development and specialization. Harold Gillies was a pioneer in the

[1] Invention Sub-Committee Meeting Minutes, 19 July 1944, File 320: Minutes of Weekly Meetings of Inventions Sub-Committee, KMG, Group 27, Box 178, DOD.

S. Chetty, *Science and Technology in South Africa, 1939-1946*, Palgrave Studies in the History of Science and Technology,
https://doi.org/10.1007/978-3-032-05084-7_6

techniques of reconstructive surgery, and he was also one of the four major reconstructive surgeons based in Britain during the Second World War—the "Big Four". The conflict foregrounded the work of another member of the Big Four, Archie McIndoe, who led the efforts made to reconstruct the burned faces and bodies of the airmen at East Grinstead. Penn was exposed to these doyens of British reconstructive surgery—and would later meet their American counterparts as well. Inspired by the work he had seen done and desirous of making his own mark in the field, Brenthurst would play a leading role in reconstructive surgery in southern Africa with imperial troops who had served in the Mediterranean and North African theatres of war among the early patients. Penn and his staff took to recording their experiences of surgery and treatment in the journal that he founded, *The Brenthurst Papers*.[2]

The journal recorded three inventions emanating from Brenthurst—the acrylic eye, the Brenthurst Splint and a "Skin Stretcher". The last, however, was not a product of the Military Hospital but was nevertheless inspired by it and was the brainchild of Alfonso Paes based at the Hospital Central Miguel Bombarda in Lourenco Marques (the present-day Mozambican capital of Maputo). Paes paid a visit to Brenthurst to observe the treatment of the war wounded. His invention however was inspired by his experiences in treating cases of "tropical ulcers" in Mozambique. Caused by bacteria, tropical ulcers or "tropical sloughing phagdena" manifests as sores developing on the unprotected legs and feet of those living in humid zones. While beginning at the level of the epidermis, the sores could extend deeper, reaching the muscle. The treatment today is usually in the form of antibiotics but also important is the debridement or cutting away of necrotic tissue, followed by the application of a skin graft.[3] For Paes, the skin graft was a necessary part of the treatment. If the wound was allowed to heal naturally, the process would be extended and the result "a large scar" accompanied by a "[weakening]" of the affected region compounded by diminished blood circulation. This

[2] Refer to Suryakanthie Chetty, *Reconstructive Surgery and Modernisation in Twentieth-Century South Africa: The Professional and Public Life of Jack Penn* (Chaim, Palgrave Macmillan, 2023).

[3] Alfonso Paes, "Presentation of a New Stretcher for Cutting Skin Grafts (Skin-stretcher Nobre Leitao)", *Brenthurst Papers,* Jack Penn (ed) (Johannesburg, Witwatersrand University Press, 1944), p10, Brn. No: 3297, Call No: 617.95 BRE, Brenthurst Library (hereafter BL) and Marie Hartley, "Tropical Ulcer", https://dermnetnz.org/topics/tropical-ulcer, accessed 18 March 2024.

would in turn make the area susceptible to "any slight trauma;" ultimately, it would never be healed fully. A skin graft was therefore necessary to permanently strengthen the site of the ulcer—and a cosmetic imperative as well.[4]

Paes and his team attempted the various types of grafts possible with the goal of "[procuring] the best method, with the simplest technic necessitating the shortest time for its execution". In addition to considering the efficacy of grafting techniques that required the least time for the operation, there was a need for suitable surgical instruments. The first was an adaptation: the use of amputation knives that were given a keen edge and then triumphantly renamed "grafting knives". Dr Nobre Leitao was one of Paes's assistants who was accredited with the development of the "skin-stretcher", an improvement of existing skin-stretchers used "to stretch the skin for the cutting of grafts". It was subsequently employed with great success by the rest of the team and considered to be so user-friendly that it could be employed by "any general surgeon".[5] It was also this accessibility that Penn would consider one of the main advantages of the Brenthurst Splint—an invention that did not require expertise but one which was nevertheless efficacious.

While Paes's work was only marginally related to the Brenthurst Military Hospital, it was nevertheless part of Penn's desire to record and publicize the reconstructive surgical innovations and refinements developed in southern Africa. The range of cases provided the "wider experience" that led to innovation and an improved way of carrying out surgeries that both exceeded the "normal boundaries of knowledge" and were also "short cuts to better results". This was related to Penn's own perception of the importance of innovative thinking in the field. As a result, due to the difficulty in publishing in the traditional northern hemisphere journals of surgery, Penn elected to establish his own journal, the *Brenthurst Papers*, the printing of which was funded by the Oppenheimers, and which would be published four times a year until 1946. According to Penn, it was the first of its kind: "the first journal devoted entirely to plastic surgery in the English language".[6] It would also be

[4] Paes, "Presentation of a New Stretcher", p10.

[5] Paes, "Presentation of a New Stretcher", pp10–11.

[6] Jack Penn, *The Right to Look Human: An Autobiography* (Johannesburg, Hugh Keartland Publishers, 1976), pp93–94.

the journal where developments in extra-oral splintage and the prosthetic eye would be published, with the former creating a dilemma as it highlighted the disjuncture between copyright, ownership and the universalist humanitarian discourse of medicine.

The Acrylic Eye

Eye injuries were a common feature of warfare in the twentieth century. The various theatres of war in which South African troops fought during the First World War provided ample opportunity for the incurring of injury, whether due to weapons or a hostile natural environment. Captain Damant of the 2nd South African Rifles was wounded in Johannesbrucke, German East Africa. He suffered "damage to the bone [and] contusion of [his right] eyeball" along with an injury to the arm which necessitated the removal of shell fragments. Sergeant E. McFarlane of the 5th South African Infantry suffered damage to his left eye in Kitanda when his machine gun was hit by a bullet, fragments of which were deflected into his eye. And the unfortunate E.J. Wilcox was blinded in the left eye when he was stung by an insect and was subsequently unable to obtain early treatment for the injury.[7]

During the interwar period then, the Directorate of Medical Services of the UDF was often involved in the obtaining and regular replacement of artificial or prosthetic eyes for veterans. In the case of Sergeant Davison formerly of the Seaforth Highlanders, a request was made for two left eyes, with the one being "cracked and roughened with use" and the spare "roughened with use" and irritating the eye socket, thus leading to "discharge from the socket". Artificial eyes were provided by a civilian firm based in Cape Town, J. Muller and Sons, Ltd, "Ophthalmic Opticians". They were sent "an old eye [to be used] as a pattern" to form the basis of the replacement. The cost ranged from 20/- for a single eye to 15/- per eye if three were ordered.[8]

The need for artificial eyes was subsequently felt early on during the Second World War due to the use of explosive "booby traps" used by the Italians to great effect among Allied soldiers. Penn described an

[7] File 104/34, Medical Boards, Chief Staff Officer (CSO), Box 63, DOD.

[8] File MD 4/5/223, Director Medical Services, Group 1, Box 47 and "Alston, CW, Lt. Col, S.A.H.A re Artificial Eyes", Director Medical Services, File MD 4/4/181, Group 1, Box 45, DOD.

"innocuous-looking fountain pen" which was a cunning explosive device that led to the loss of hands and eyes of the innocent user. The damaged eye had to be surgically excised, and the early procedure used in England was the subsequent insertion of a "gutta-percha ball"[9] to serve as a mould which was then overlain with a skin graft. The eyelids were then sutured shut to hold it all in place. This was done to prevent the eye socket from shrinking. After three months, an artificial eye was inserted into the socket.[10] Penn's first intervention came in the form of the "acrylic 'gun turret' mould", a device using acrylic that had been shaped into the approximation of the shape of an artificial eye. Inserted at angles into the mould were two steel tubes giving the device its "gun turret appearance". The tubes functioned as a system of lubrication and drainage, with a saline solution passed through one of the tubes via a syringe and the detritus emerging from the other tube. The regular system flushed out impurities and kept the socket clean thus reducing the chances of sepsis that were a feature of the gutta-percha ball. Moreover, it significantly reduced recovery time and the afflicted was ready to be fitted with an artificial eye after only five days. The time spent in hospital was minimal and the patient was instructed in "correct orbital hygiene" and told to wear the artificial eye more or less continuously for the ensuing four months.[11]

The second aspect of the operation was of course, the development of the acrylic eye. Glass had been the preferred material for prosthetic eyes since the sixteenth century, with Venice—with its illustrious history of glass blowing—an early centre for artificial eye production. Glass eyes were produced in other European states as well and, by the nineteenth century, Germany had taken the lead in manufacturing glass eyes. It was also during this time that the Germans attempted custom-made glass eyes

[9] Gutta-Percha is a substance derived from the Palaquium gutta tree which is indigenous to South East Asia. With similar properties to rubber, it was first used in medicine in the mid-nineteenth century for splints and the handles of medical implements as well as in the stemming of bleeding and for skin ailments. It was also used as a filling for root canals. Refer to R. Prakash, V. Gopikrishna and D. Kandaswamy, "Gutta-Percha: An Untold Story", *Endodontology*, 17(2), 32–36, 2005, pp32–33.

[10] Penn, *The Right to Look Human*, p88.

[11] Major Jack Penn, Major L.J. Brown, Captain T.B. Berry, Sgt. A. Schulmeister, Sgt. J. Roylance and Mrs C.L. Ormond, "Problems in Eye Lid and Socket Reconstruction", *Brenthurst Papers*, Jack Penn (ed) (Johannesburg, Witwatersrand University Press, 1944), p10, Brn. No: 3297, Call No: 617.95 BRE, BL, pp11–12.

at the Leipzig Eye Institute. Very brittle glass was soon replaced by cryolite glass, giving it the greater durability of the mineral. By the beginning of the twentieth century—and before two world wars—the British alone bought fifty thousand glass eyes annually at £1 each, and these would have to be replaced every year.[12] Two years before the outbreak of the First World War, Germany was producing and exporting 180,000 kilograms of optical glass, with a quarter of that going to the United States and to Britain. With the First World War, these exports ceased just as the combination of artillery and lack of protective gear increased eye injuries and stimulated demand for prosthetic eyes.[13] This scenario would be repeated during the Second World War.

With the Nazi assumption of power in 1933, Germany banned certain materials exported to Britain, among them the glass used in glass eye manufacture, creating a shortage. With the outbreak of war and the growing demand for prosthetic eyes, attempts were made to manufacture eyes from other materials and both the British and the Americans turned to the use of an acrylic, Poly (methyl methacrylate).[14] Jack Penn would however claim the primacy of invention for Brenthurst.

South Africa relied on the importation of glass eyes, and the disruption to supply lines and shipping caused by the War resulted in a shortage in South Africa. Further, according to Penn, there were no glass blowers in South Africa able to take on this skilled work. While acrylic would prove to be an improvement over glass, its development as an artificial eye was due to the exigencies of war and the first artificial acrylic eyes were made in 1941, which Penn claimed to be the first of their kind ever used in reconstructive surgery. Plastic and metal had been used with increasing frequency during the War to substitute for non-reactive celluloid and organic material such as ivory in implants. Penn, initially leery of the use of these materials, would eventually, and successfully, use an acrylic plate on a patient with a severe injury that had exposed his brain.[15]

[12] Keith Pine, Brian Sloan and Robert Jacobs, "A Brief History of Prosthetic Eyes", https://eyemuseum.ranzco.edu/wp-content/uploads/2019/06/History_V4_2.pdf, accessed 20 March 2024.

[13] Evan P. Sullivan, "The Eye at War: American Eye Prosthetics During the World Wars", 25 September 2017, *Nursing Clio*, https://nursingclio.org/2017/09/25/the-eye-at-war-american-eye-prosthetics-during-the-world-wars/, accessed 20 March 2024.

[14] Pine, et al., "A Brief History of Prosthetic Eyes".

[15] Penn, *The Right to Look Human*, pp88–89, 91.

The acrylic eye, while revolutionary, thus presented less of a break with existing treatment and more a continuation of an ongoing trend.

Three figures are credited with the development of the acrylic eye—Sergeants Schulmeister and Roylance who were dental technicians at Brenthurst and Cecile Ormerod. While the sergeants focused on the more technical aspects of the eye such as its shaping, Ormerod was largely responsible for its "natural" appearance. Penn was a sculptor of note and placed value on the visual aspect of reconstructive surgery—it was a combination of science and art. Cecile Ormerod had therefore been hired as an artist at Brenthurst to address the aesthetic features of reconstructive surgery.[16]

The procedure for the implanting of an artificial eye began with the making of a wax mould of the eye socket. Drawing upon the methods of creating dental moulds, the eye was then cast in acrylic, polished and placed in the eye socket for two weeks so that the eye socket became "[accustomed]...to its shape". An iris was painted on paper and this was varnished onto the mould. The colour of the iris was based on comparisons with the existing eye and the pupil size was based on the larger pupil size indoors instead of the contracting pupil in bright light, as the former was likely to "[attract] the least attention". The emphasis on rendering the eye as natural as possible was evident in the use of pink silk thread, also varnished onto the eye, that resembled blood vessels. The final product was covered with clear acrylic and polished.[17]

From the outset, the properties of the acrylic eye were manifestly superior to its glass counterpart: it was lighter; being individually rather than mass produced meant that it could be shaped for any socket; and it was "unbreakable".[18] Penn tested the latter property in a public "exhibition of showmanship" by flinging a sample against a wall whereupon it bounced around the room, emerging with nary a scratch. Needless to say, glass eyes were far more brittle and susceptible to damage which became incredibly inconvenient if the soldier was stationed in a remote area.[19] It was the notion of convenience and of utility, particularly in war, that would apply to the other major invention emanating from Brenthurst—the splint.

[16] Penn, *The Right to Look Human*, p77.

[17] Penn, et al., "Problems in Eye Lid and Socket Reconstruction", p16.

[18] Penn, et al., "Problems in Eye Lid and Socket Reconstruction", p16.

[19] Penn, *The Right to Look Human*, p89.

The Brenthurst Splint was a utilitarian and pragmatic invention, based on the identification of a specific need. It was based on a collaborative effort between Penn and dental surgeon Lester Brown. In their account of the development of the device, Penn and Brown addressed the shortcomings of existing forms of splintage used to immobilize the jaw in the event of fracture, finding that the methods were invasive with a greater potential for infection. Moreover, they were not suitable for all forms of injury. Contemporary reconstructive and dental surgeons, including Gillies, his dental surgeon associate William Kelsey Fry and another member of the pioneering "Big Four", Arthur Rainsford Mowlem, employed the use of Kirschner wires on the displaced "fragments" of the jaw that were joined with a "cross bar" to provide stability.[20] Arthur Mowlem was based at Hill End Hospital in Hertfordshire, becoming part of the "Big Four" in 1939. His work at Hill End provided opportunity for research and investigation, and he was tasked by Alexander Fleming to test out penicillin which was given to patients orally, intravenously and as a topical treatment. Mowlem, along with dental surgeons Benjamin Fickling and Alexander MacGregor, was also a pioneer in the use of steel pins to immobilize fractured jaws.[21] In 1941, Mowlem and his colleagues at Hill End wrote an article describing the use of an "external pin" method for the stabilization and immobilization of fractured jaw—which Penn and Brown referenced. Integral to the method was the use of Kirschner cross wires. Developed by Martin Kirschner in 1909, what became known as Kirschner or K-wires, were initially metal pins that would eventually become thin, stainless steel wires. While the metal pin was useful for "thick long bones"—and used in orthopaedic surgery—K-wires lent themselves to smaller and more delicate bones such as those found in the jaw and, where they are primarily used today, the hand. The initial pins had also required the bone to be drilled first before they were inserted—which increased the rate of infection. In the interwar period, Kirschner also developed a method for inserting the steel wire directly through the skin and into the bone. New methods of insertion, the reduced diameter of the wire as well as

[20] Major Jack Penn and Captain Lester Brown, "A New Conception in Extra-Oral Splintage", *Brenthurst Papers*, Jack Penn (ed) (Johannesburg, Witwatersrand University Press, 1944), p10, Brn. No: 3297, Call No: 617.95 BRE, BL, pp3–4.

[21] R.L.G. Dawson, "The History, Antecedents and Progress of the Mount Vernon Centre for Plastic Surgery and Jaw Injuries, Northwood, Middlesex 1939–1983", *British Journal of Plastic Surgery*, 41, 83–91, 1988, pp84–85.

the development of antibiotics, meant that K-wires were widely used by surgeons during the Second World War.[22]

Mowlem's device consisted of two sets of two pins (the wire) inserted into either side of the break in the bone. The wires were inserted at an angle of 60 degrees to each other forming a cross and each pair was held in place with a "locking plate" keeping them in position. The two locking plates were then connected by a "fixation bar". The wires were approximately 5–6.5 centimetres in length tipped with a "short bevelled pyramidal point" and needed to be at least 1.5 millimetres in diameter for stability. While acknowledging that the insertion of the wire of the required thickness through the face may be construed as "an unwarranted barbarism", Mowlem nevertheless stressed that the pain experienced by the patient would be minimal, the device allowed for greater flexibility of movement between the upper and lower jaw allowing the patient "eat in comfort" and to maintain "oral cleanliness". Further, the slight scars due to the wire could be removed surgically.[23] These were all features that the Brenthurst Splint was designed to emulate and to improve.

Penn had experienced the methods used by Mowlem et al. during the brief period he spent in Britain as one of two members of the UDF sent to learn the rudimentary skills of plastic surgery under the "Big Four". He was convinced however that the use of the pins inserted into bone often had unforeseen and dangerous side effects such as penetrating of the nerves, the "floor of the mouth" and even osteomyelitis, the infection of the bone. He therefore sought to find a solution that could maintain the positioning of the broken jaw while minimizing invasive contact with the bone itself. For inspiration he turned to the past, investigating the collection of jaw bones held at the Department of Anatomy at the University of the Witwatersrand. These bones were of those who had died at various ages so could demonstrate change in the bone structure over time. With the discovery of early hominins in South Africa from 1924, Penn also looked at fossil specimens. Having studied and worked at the University

[22] Bas B.G.M. Franssen, Arnold H. Schuurman, Aebele Mink Van Der Molen and Moshe Kon, "One Century of Kirschner Wires and Kirschner Wire Insertion Techniques: A Historical Review", *Acta Orthopaedica Belgica*, 76, 1–6, 2010, pp1–2.

[23] Rainsford Mowlem, J.L. Dudley Buxton, Alexander B. MacGregor, John N. Barron, "External Pin Fixation for Fractures of the Mandible", *The Lancet*, 4 October 1941, pp392–393.

of the Witwatersrand, Penn had provided the illustrations for the palaeonthropological papers written by Raymond Dart and his colleagues and felt well placed to interpret the fossils.[24]

Penn's conclusion was that the feature shared by all the mandibles he examined was that the "lower border" remained the same. Above this was the teeth—the number and condition of which varied among individuals. It was the region containing the teeth where most jaw injuries were sustained, whether as a result of car and aeroplane crashes or due to violence, but the border below it tended to remain "constant". Penn's solution then was the use of "claws" or "hands" that could be applied extra-orally that would hold the border in position as the jaw healed. This was the plan he laid before Lester Brown with Penn focusing on the "claw" aspect and Brown experimenting with the "universal joints and cross bars" that would hold the device together.[25]

In developing their early models—of which there would be twenty-two—Penn and Brown sought the assistance of the skilled metallurgists and engineers of the South African Railways and Harbours Administration (SAR&H).[26] The mechanics and engineers of the SAR&H were a significant part of South Africa's war effort. In 1941, the Railways and Harbours Base Tank Workshop was sent to the Middle Eastern theatre. After a twenty-six day voyage, they arrived at their destination in the midst of an air-raid and within a day had begun their work of repairing and maintaining Allied tanks.[27] Along with work done on aeroplanes, tanks and other military vehicles the SAR&H also played a significant role in the design and development of instruments used for medical treatment. The Mechanical Engineer's Department collaborated with the Department of Defence to produce much-needed orthopaedic tables, surgical tables that could be adjusted for the dimensions of the individual and could accommodate the taking of X-rays.[28] Another device produced by

[24] Penn, *The Right to Look Human*, pp82–83.

[25] Penn, *The Right to Look Human*, p83.

[26] Penn, *The Right to Look Human*, p83.

[27] "R & H Base Tank Workshop: Technical Services Corps", *South African Railways and Harbours Magazine*, Vol 36, May 1942, pp415–416, https://railways.haarhoff.co.za/, accessed 21 February 2024.

[28] "Orthopaedic Tables Made by Mechanical Department", *South African Railways and Harbours Magazine*, Vol 36, February 1942, pp119–120, https://railways.haarhoff.co.za/, accessed 21 February 2024.

the workshops in the South African capital, Pretoria, were hospital trolleys that needed to be "light but sturdy" and minimize the "jarring" to patients. Fifteen were subsequently made to be used in base hospitals.[29] Working alongside inventors to render ideas feasible and practical, the SAR&H were the ideal collaborators in realizing Penn and Brown's design.

The Brenthurst Splint consisted of three features: the clamps that would fasten to the jaw and hold it in position; the universal joints such as the locking plate that would link these; and connecting bars to hold the device together. Numerous models were tried and tested and by the time it was complete, the splint was designed for "the ease of application". The various components that comprised the splint also came with a custom spanner and key for assembly.[30] Even after the splint had received some publicity—resulting in both domestic and international demand—there were delays in meeting the demand. As the Electrical Workshops of the SAR&H reported to the Director General of Medical Services (DGMS), the metal that composed the components was of uneven hardness which presented challenges for drilling as well as "tapping"—or the cutting of threads. The latter required specially adapted tools.[31] Eight months later, there were still some "minor flaws" in the production process.[32] Penn considered these "flaws" easily remedied: the "screw pins" were not long enough to reach the bone and the "connecting hooks" did not attach to the joints as effectively as they should. He recommended the recall of the boxes of splints already sent out and their replacement by the perfected versions and was particularly concerned that the recipients of the splints sent abroad needed to be informed of these flaws and be "[assured]" that they would soon be in possession of "efficient splints". For the perfectionist Penn, any flaw detracted from the "fool-proof" nature of

[29] "Another Contribution by the S.A.R. & H. Mechanical Department to the War Effort", *South African Railways and Harbours Magazine*, Vol 36, April 1942, p333, https://railways.haarhoff.co.za/, accessed 21 February 2024.

[30] Penn and Brown, "A New Conception in Extra-Oral Splintage", pp7–8.

[31] A. Everett, "re: Brenthurst Clamp Splints", 14 September 1943, File MD 540/6 MVS 615/42: "Brenthurst Splints", GG, Group 2, Box 1101, DOD.

[32] J.M. Watt, DGMS, "Brenthurst Splints", 12 May 1944, File MD 540/6 MVS 615/42: "Brenthurst Splints", GG, Group 2, Box 1101, DOD.

the splint and its components.[33] Further, clearly as a South African—and Brenthurst—invention, its efficacy was linked to his reputation in the international community of reconstructive surgeons. And it was touted as being an improvement on existing splints.

As with Mowlem's method, when assembled, the clamps would be positioned on both sides of the break and connected with the various joints and connecting bars. The key difference was that no drilling or insertion of pins into the bone was required, and the procedure was therefore non-invasive. The procedure for the use of the splint was relatively simple. The patient was anaesthetized, and cuts were made to the face that reached the lower border of the mandible. Each incision was approximately 1.2 centimetres in length. Once the appropriate depth had been reached, the clamps were applied and secured to the mandible. The incisions could then be stitched. The joints were then connected once the positioning had been checked. Once the jaw had healed, the patient was once again anaesthetized and the procedure was reversed—the joints and bars were removed and subsequently, the clamps. Further stitches were required with minimal facial scarring.[34]

The splint could be applied to diverse cases. Within the context of the War, Penn saw it as making a significant contribution in stabilizing the jaw of the wounded near the frontlines and before they could be evacuated to base hospitals. The same applied to those injured aboard ship and the simplicity of its use meant that it could be applied by those not trained in surgery. Simultaneously, the adjustability of the clamp system and the focus only on the inferior border meant that it could just as easily be used on children. Further, it had the tremendous advantage of reducing the rate of infection—and could, in fact, be used when infection was already present; there was little chance of damage to the bone, teeth or nerves as was the case with more invasive methods; and the patient could immediately resume a normal diet with Penn describing a case where a patient actually gained in excess of five kilograms during the healing period. This was in stark contrast to invasive intra-oral splintage that wired the jaw

[33] Jack Penn, "Re: Brenthurst Clamp Splints", 2 May 1944, File MD 540/6 MVS 615/42: "Brenthurst Splints", GG, Group 2, Box 1101, DOD.

[34] Penn and Brown, "A New Conception in Extra-Oral Splintage", pp5, 9.

shut, leading to a restricted and often liquid diet and resulting in weight loss of up to twenty kilograms.[35]

Over the course of the War, the splint was a ubiquitous device at the Brenthurst Military Hospital with more than two hundred patients treated by 1945. While some were combat or work-related, many injuries were also incurred as a result of fights breaking out between men stationed in the Union. These injuries also allowed for the refining of the splint. In the case of a patient admitted three days after an assault, the splint was applied but when it was removed, it was discovered that there was a reaction of the skin in the region of the clamp where "the bars had been soldered to the clamp". Later clamps were subsequently not soldered. In another case, the "extra-oral splint pin" (that pierced the skin) "eroded", inflaming the tissue. The splint would thereafter be manufactured using only "high grade stainless steel". There was thus no further indication of erosion or inflammation. Another patient sustained numerous fractures to the jaw when his face was caught in the machinery while working. He would require the application of five clamps across the entire jaw yet was able to continue his diet as normal. A patient admitted to the hospital three weeks after the incursion of severe injury was operated on, the clamps applied and only remained in hospital for a further three days. By all accounts, the Brenthurst Splint was a success; Penn and Brown described it as "the development of a new principle which we think has come to stay".[36] Heralded by the South African press as "revolutionary" with the promise of "immediate succour" for wounded soldiers—in contrast to British and American devices—the Brenthurst Splint exemplified national ingenuity and a significant contribution to the war effort.[37]

In July 1944, Penn wrote to the Director General Technical Services (DGTS) describing the two inventions made at Brenthurst and enclosed a copy of the *Brenthurst Papers* that detailed their production and use. He requested a "gratuity" for those involved in the development of the acrylic

[35] Penn and Brown, "A New Conception in Extra-Oral Splintage", pp5, 12 and Penn, *The Right to Look Human*, p83.

[36] Penn, *The Right to Look Human*, p83 and Penn and Brown, "A New Conception in Extra-Oral Splintage", pp12–17.

[37] "City Dentists Invent Revolutionary Surgical Appliance", *Rand Daily Mail*, 1 April 1943.

eye—Sergeants Schulmeister and Roylance and Mrs C.L. Ormerod—which was already in use by the UDF. Demonstrating a disarming lack of proprietorship, it was evident that neither he nor Lester Brown had taken out a patent for the Brenthurst Splint and Penn further suggested that, due to its proven efficacy in the testing phase, the UDF may desire to "[obtain]" it.[38]

A follow-up letter to the DGTS raised the issue of compensation and Penn made a case for the invaluable significance of the two inventions to substantiate his request. The acrylic eye was an improvement on existing glass eyes: it could be custom designed and made for the user, was a more flexible and forgiving material than glass and was "virtually everlasting". The Brenthurst Clamp Splint had its own advantages: it could be used on the frontlines or on ships where there was little access to hospitals, could be used on the most challenging jaw fractures and allowed the wearer to eat and function normally. Both then, were a definite improvement on existing devices. Yet Penn initially made clear that, despite the worth of these inventions, those involved in their creation were not seeking monetary reward as this could possibly increase the cost of the "appliances", thus making them unaffordable for those who may have need of them. He did request that the inventors be given the funding for a trip to the United States as "an SAMC [South African Medical Corps Team] representing South African plastic and maxillo-facial surgery". The rationale for the trip was the sharing of knowledge, the opportunity to acquaint American medical personnel with the use of the devices but also to showcase the developments at Brenthurst and by so doing, contribute to the "SAMC war effort…[and] bring us credit".[39] A subsequent letter altered the initial disregard for monetary compensation when Penn informed the DGTS that Sergeants Schulmeister and Roylance "were rather 'hard up' at the moment" and requested a "cash gratuity, rather than take advantage of a course overseas".[40] There is a slight censorial hint of disapproval

[38] Jack Penn, "R.O. 1057 – Inventions", 5 July 1944, File 332/1: 1) Brenthurst Clamp Splint 2) The Acrylic Artificial Eye—Brenthurst Inventions, KMG, Group 27, Box 163, DOD.

[39] Jack Penn, 11 July 1944, File 332/1: 1) Brenthurst Clamp Splint 2) The Acrylic Artificial Eye—Brenthurst Inventions, KMG, Group 27, Box 163, DOD.

[40] Jack Penn, 13 July 1944, File 332/1: 1) Brenthurst Clamp Splint 2) The Acrylic Artificial Eye—Brenthurst Inventions, KMG, Group 27, Box 163, DOD.

in Penn's tone at the subordination of education to lucre but he nevertheless advocated that due recognition and compensation be given to all figures involved in the process of invention.

The three points raised by Penn were considered by the office of the DGMS whose conclusions were emphatically made with a liberal use of exclamation points. The request for a gratuity for the sergeants was accorded a "Yes!" with the recommendation that they be paid £25 each.[41] This was subsequently raised to £50 each.[42] Unfortunately, for Penn, the request for the trip was given just as resounding a "No!" with the reasoning being that it was neither in the "national interest" nor was it "the time for such visits".[43] Penn's hope then to use inventions to assert South African developments in reconstructive surgery was dashed. The final point addressed by the DGMS however related to the patenting of the Brenthurst Splint and here, again, the response was an unequivocal "No!" For the DGMS, there was no need to patent the device unless it could potentially be used as—or altered to be—a weapon. Significantly, it was a device used for healing and the DGMS did "not consider that patenting these devices is in the spirit and traditions of the medical profession, and cannot therefore recommend it".[44]

Penn (and Brown) had waived their patent rights to the Brenthurst Splint in favour of the UDF.[45] Penn, however, had already published comprehensive details about the development, manufacturing and utilization of the splint in the *Brenthurst Papers*, the copyright of which was held by him. Further, as evident by his request to demonstrate the splint in the United States, Penn was eager to share the device with other reconstructive surgeons in the Allied countries. Other than its value in treating

[41] D.M. Watt, DGMS, "Brenthurst Clamp Splint and Acrylic Artificial Eye", 16 August 1944, File 332/1: 1) Brenthurst Clamp Splint 2) The Acrylic Artificial Eye—Brenthurst Inventions, KMG, Group 27, Box 163, DOD.

[42] Secretary, Inventions Committee, "Brenthurst Clamp Splint Acrylic Artificial Eye", 29 August 1944, File 332/1: 1) Brenthurst Clamp Splint 2) The Acrylic Artificial Eye—Brenthurst Inventions, KMG, Group 27, Box 163, DOD.

[43] D.M. Watt, DGMS, "Brenthurst Clamp Splint and Acrylic Artificial Eye".

[44] Watt, DGMS, "Brenthurst Clamp Splint and Acrylic Artificial Eye". It should be noted that August 1944 was a pivotal point in the War with the liberation of France after the massive Allied invasion of Normandy.

[45] DGMS, "Brenthurst Clamp Splints", 9 January 1947, File 332/1: 1) Brenthurst Clamp Splint 2) The Acrylic Artificial Eye—Brenthurst Inventions, KMG, Group 27, Box 163, DOD.

injuries to the jaw, the splint (like the journal, the *Brenthurst Papers*) was a means of demonstrating South African prowess and progress in the field of reconstructive surgery, as evident at Brenthurst. While the UDF had set the price of the splint and its associated toolkit at £60, sets were gifted to various branches of the British Armed Services. Penn also requested that sets be sent to the National Committee of Plastic and Maxillo-Facial Surgery in the United States, the Surgeons General of the United States Army and Navy, the Director General of the Indian Army Medical Service and the Director General of the Medical Services of the USSR. Clearly uneasy about the unauthorized replication of the splint, the DGTS recommended that copyright be established in all the countries to which the splint would be sent so as to "act as a deterrent against plagiarism".[46] While acknowledged as a medical device for humanitarian and healing purposes, the Brenthurst Splint demonstrated concerns regarding proprietorship and the sharing of knowledge and technology between the Allies (Fig. 6.1).

A Special Relationship Between Allies

As discussed in Chapter Two, agreements were made between Allied countries related to the sharing of technology and the subordination of patent rights to the war effort. Further, enemy technology—with many German inventions during the interwar period adapted and utilized by the Allies—was commandeered for the War with no recognition of the rights of inventors. As such, the sharing and use of technology (and inventions) became an expression of international alliances and enmities. Yet, even among Allied nations, the sharing of technology was an intricate process, marked by national, ideological and security concerns.

While American bombers of the United States Air Force used the Norden bombsight to great effect (if not always great accuracy) during the bombing campaigns over Germany, the bombsight was also an early example of the ostensible American maintenance of neutrality. The Norden M-9 bombsight was an early computerized targeting system invented by Carl Norden to improve the accuracy of bombing. The bombardier would enter the direction of the wind as well as the speed

[46] C.H. Blaine, Secretary for Defence, 13 November 1944, File 332/1: 1) Brenthurst Clamp Splint 2) The Acrylic Artificial Eye—Brenthurst Inventions, KMG, Group 27, Box 163, DOD.

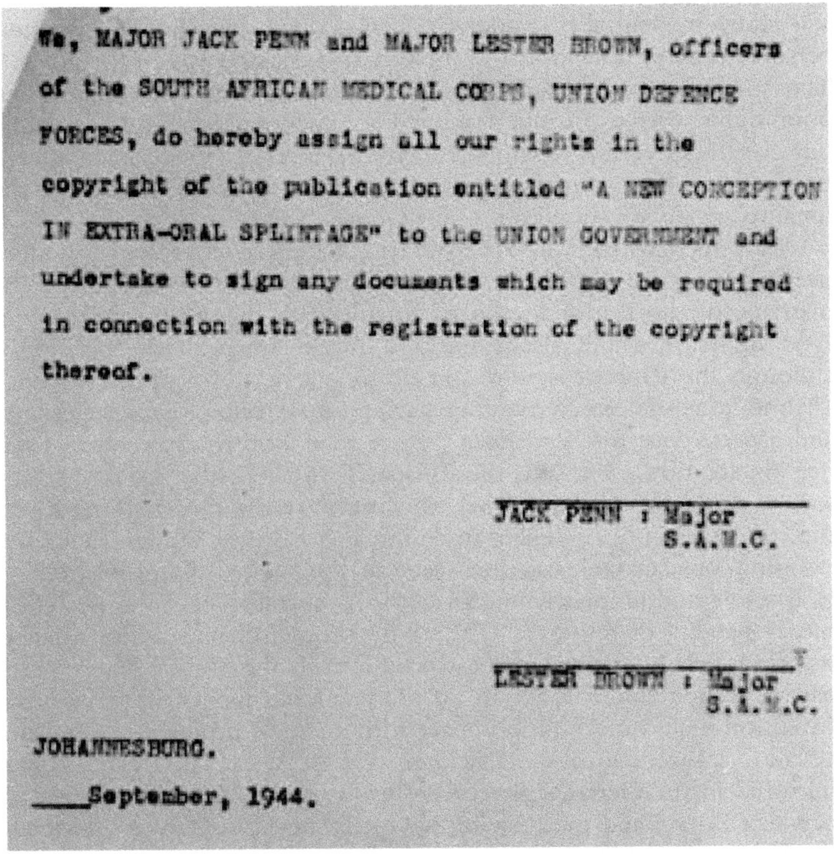

We, MAJOR JACK PENN and MAJOR LESTER BROWN, officers of the SOUTH AFRICAN MEDICAL CORPS, UNION DEFENCE FORCES, do hereby assign all our rights in the copyright of the publication entitled "A NEW CONCEPTION IN EXTRA-ORAL SPLINTAGE" to the UNION GOVERNMENT and undertake to sign any documents which may be required in connection with the registration of the copyright thereof.

JACK PENN : Major
S.A.M.C.

LESTER BROWN : Major
S.A.M.C.

JOHANNESBURG.

____September, 1944.

Fig. 6.1 Brenthurst Splint agreement, taken from M.D. 540/6 MVS 615/42: "Brenthurst Splints", GG, Group 2, Box 1101, DOD

and altitude of the bomber and based on these calculations, the correct trajectory to the target was determined. It was considered revolutionary in terms of its accuracy—and certainly an improvement on existing models.[47] During the tense days of August 1939, when the United States

[47] "Norden M-9 Bombsight", National Museum of the United States Air Force, https://www.nationalmuseum.af.mil/Visit/Museum-Exhibits/Fact-Sheets/Display/Article/196340/norden-m-9-bombsight/, accessed 1 March 2024.

still clung to neutrality, Roosevelt turned down Neville Chamberlain's personal request for Norden bombsights to be used in British planes.[48] The Americans would eventually assemble some ninety thousand Norden bombsights, even as its much vaunted accuracy was called into question, however the perceived advantage of the bombsight would be countered by the time the Americans were drawn into the conflict when a spy obtained its design specifications for German use.[49]

With American entry into the War, the sharing of technology was more overt and marked by collaboration—yet national interests remained. The marriage of the British Merlin 61 engine with the American Mustang P-51 fighter exemplified the spirit of Anglo-American co-operation—although the Americans were initially reluctant to adopt the superior "hybrid" due to deep conviction regarding their technological superiority and a preference for "American [rather than British] channels of engineering scrutiny". Similarly, the Russian T-34 tank—that would become such a thorn in the side of German forces on the Eastern Front—was the combination of an American design and a British Vickers Tank. The resulting Russian tank was then sent to the United States in 1942 to be assessed and ultimately improved.[50] Co-operation was also related to the availability of resources. The cavity magnetron radar, for instance, was invented in Britain but mass produced in the United States which had the production capacity to do so. The same applied to penicillin.[51] Invention would make little impact without mass production—and not all countries were equal—and neither were all Allies. This inequality was mirrored in the hierarchical relationship between Allied countries—and between Britain and the Dominions. As discussed in Chapters Two and Four, where South Africa lacked the scope to develop an idea and to test it, it was forwarded to Britain or the United States.

In the case of the Brenthurst Splint, invention did not operate in a vacuum and a request was made to have two boxes sent to the Royal

[48] Richard Overy, *1939: Countdown to War* (Penguin, 2010), p33.

[49] "Norden Bombsight", *Air and Space Magazine*, April 2020, https://www.smithsonianmag.com/air-space-magazine/norden-bombsight-180974472/, accessed 1 March 2024.

[50] Paul Kennedy, *Engineers of Victory: The Problem Solvers Who Turned the Tide in the Second World War*, (Penguin, 2014), pp121–123, 187–191.

[51] Tom Shachtman, *Laboratory Warriors: How Allied Science and Technology Tipped the Balance in World War II* (New York, Perennial, 2003), pp129, 167–169.

Air Force "Facio-Maxillary centres" so that comparisons could be made between their methods of splintage and that presented by the splint. The RAF Medical Liaison Officer stationed at the DGMS believed that the efficacy of the methods employed could only be obtained if direct comparisons could be made and that, based on the spirit of co-operation between the Empire's "various Air Force Medical Services…the gift of these two sets would be deemed an impressive gesture". The DGMS subsequently authorized the "gift" free of charge.[52] In return, UK practitioners were expected to test the device and subsequently submit their "observations on [its] utility".[53]

With Penn and Brown's publication of the account of the development of the splint in the *Brenthurst Papers*, international interest was aroused—and Penn was keen to capitalize on this. In a New Year's missive to the DGMS, he requested the sending of boxes of splints to a number of figures and organizations in preparation for the fighting that was anticipated in the Pacific theatre of operations but also to T. Pomfret Kilner at the Pensions Hospital in Roehampton as well as to the R.N. Hospital in Ceylon [Sri Lanka] who had urgent need of the splints. Penn, moved by a humanitarian impulse, believed it would "be a good gesture on our part" that would not affect adversely the availability of the splints for domestic use.[54] The response by Lt Colonel Watt of the DGMS acquiesced to Penn's request but the letter suggested that Penn's demands conflicted with the pecuniary and proprietary interests of the UDF. Added to the comment, "Kindly inform him that steps are being taken as requested in his minute", was the handwritten, "but that it is impossible to consider wholesale gifts in this way".[55]

Kilner's appearance as a recipient of the "gift" is a significant one. Thomas Kilner was one of the "Big Four", and the only one to be born in England. During the First World War, he worked alongside Harold Gillies at Sidcup and, at the end of the War, both men could be considered

[52] J.M. Watt, DGMS, "Gift of Brenthurst Splint", 26 October 1943, File MD 540/6 MVS 615/42: "Brenthurst Splints", GG, Group 2, Box 1101, DOD.

[53] J.M. Watt, DGMS, 10 November 1943, File MD 540/6 MVS 615/42: "Brenthurst Splints", GG, Group 2, Box 1101, DOD.

[54] Jack Penn, "Re Distribution of Brenthurst Clamp Splint", 1 January 1944, File MD 540/6 MVS 615/42: "Brenthurst Splints", GG, Group 2, Box 1101, DOD.

[55] J.M. Watt, DGMS, "Disposal of Brenthurst Splints", 5 January 1944, File MD. 540/6 MVS 615/42: "Brenthurst Splints", GG, Group 2, Box 1101, DOD.

the only reconstructive surgeons based in Britain. During the interwar period and after, he sought to rectify this by playing a significant role in furthering the teaching and training of surgeons and establishing reconstructive surgery as a distinct field of specialization. During the War itself, he was stationed at the Stoke Mandeville Hospital and also led the plastic surgery unit at the Ministry of Pensions.[56] Of the four reconstructive surgical pioneers, Penn would describe Kilner as "the man he liked best".[57]

Sending a "gift" to another member of the Big Four aroused the ire of the UDF. In the wake of Penn's largesse—despite his disregard for financial considerations—he was subsequently asked to explain the distribution of two splints to a Major Champion of the No 1. Maxillae Facial Unit, Middle Eastern Forces, and to Harold Gillies himself. The request for an explanation came with a stringent warning, "Please note that under no circumstances whatsoever will you issue any articles of Defence property without the authority of or reference to this office".[58] The co-inventor of the Brenthurst Splint found himself having to explain that the two boxes were "the first made by the [SAR&H]" that had been given to him to check before they went into production.[59] As discussed in the previous chapter, A.H.R. Champion was a reconstructive surgeon in the Royal Army Medical Corps who had inspected the various reconstructive surgery facilities in South Africa in 1942 with particular attention paid to Brenthurst and its technological innovations: the acrylic eye and the splint. Harold Gillies was, of course, the pioneering figure in modern reconstructive surgery since the First World War. In this case then, Penn's distribution of the splint (*sans* permission) was less a humanitarian gesture than it was an attempt to showcase Brenthurst inventive prowess.

Other seminal figures in reconstructive surgery were also to be recipients of the splint. A month later, Penn re-iterated his request for a box to be sent to Kilner. Another box was to be sent to Robert Ivy in the United States. Penn felt it would be "beneficial to our Medical Corps"

[56] Kilner, Thomas Pomfret (1890 - 1964), *Plarr's Lives of the Fellows*, https://livesonline.rcseng.ac.uk/client/en, accessed 25 March 2024.

[57] Penn, *The Right to Look Human*, p71.

[58] DDMS Witwatersrand Command, "re: Brenthurst Clamp Splint", 11 April 1944, File MD 540/6 MVS 615/42: "Brenthurst Splints", GG, Group 2, Box 1101, DOD.

[59] Jack Penn, "Re Brenthurst Clamp Splint", 13 April 1944, File MD 540/6 MVS 615/42: "Brenthurst Splints", GG, Group 2, Box 1101, DOD.

if these two men were given access to the splint and that he would be willing to purchase two boxes and forward it to them if the UDF did not see fit to send them complimentary boxes. In his letter Penn described Ivy—who was another key figure specializing in reconstructive surgery, serving as the Professor of Plastic Surgery at the University of Pennsylvania and "Chairman of the United States National Committee for Plastic and Maxillo-Facial Surgery"—as playing a central role in the development of "extra-oral splintage" who was much taken with the "descriptions" provided of the splint and had thus requested a set.[60] Penn was an admirer of Ivy and in particular, the Ivy Plan that provided for reconstructive surgery to be performed on children born with facial congenital defects such as the cleft lip and palate.[61] While subordinate to the dictates of the UDF then, Penn nevertheless perceived the splint as a means of creating and maintaining international ties with the pioneering and leading specialists in his field and there was a clear tension between his vision for the splint and that of the UDF.

The international distribution of the splint also took on an additional complexity when the splint was sent to the Soviet Union as part of a humanitarian gesture by the South African-based "Medical Aid for Russia". After one set was sent to the Moscow Traumatological Institute, the Soviets requested an additional fifteen sets. Writing to the DGMS, the secretary for "Medical Aid for Russia" stressed the "enormous casualties" incurred by the Soviets, and hence the need for the splint.[62] The UDF however only had ten sets available at a total cost of £600 which was duly paid by the organization.[63] An ostensibly simple transaction, it took on new significance in light of Penn's ardent anti-communism.

As Beardsley discusses, the inclusion of the Soviet Union in the Allied sharing of technology was not uncomplicated. From 1942, the United

[60] Jack Penn, 15 May 1944, File MD 540/6 MVS 615/42: "Brenthurst Splints", GG, Group 2, Box 1101, DOD.

[61] Penn, *The Right to Look Human*, p101; Mansher Singh, R. Barrett Noone, Halimat Afolabi and Edward J. Caterson, "A Tribute to the Founding Figures of the American Board of Plastic Surgery", *Plastic and Reconstructive Surgery*, 135, 6, 1047–1054, June 2015, pp1049–1050.

[62] M. Schneier, Medical Aid for Russia, 19 December 1944, File MD 540/6 MVS 615/42: "Brenthurst Splints", GG, Group 2, Box 1101, DOD.

[63] M. Schneier, Medical Aid for Russia, 4 January 1945, File MD 540/6 MVS 615/42: "Brenthurst Splints", GG, Group 2, Box 1101, DOD.

States and Britain vacillated over the sharing of "technical secrets" with the Soviet Union with each either supporting or opposing collaboration over the course of the War. In 1942, for instance, the West's sharing of technology with the Soviet Union was a means of taking the sting out of the Allies' inability to come to the defence of the Soviet Union as it fought against the German invaders. An agreement was reached between the British and the Soviets regarding the sharing of information "for the prosecution of the war against the common enemy". In contrast, the Americans were wary, fearing that while the Soviets would benefit from the agreement, it would do little for the Americans. The situation was rendered even more difficult by the British privileging their alliance with the Americans over that with the USSR. The Americans thus checked the 1942 accord, forbidding the sharing of information with the Soviets related to guided missiles, radar, jet engines and other technology; the Soviet should only be allowed data that they already held. By 1944, realizing the importance of Soviet resources in winning the War in the Pacific—which was expected to drag on for another three years—the Americans adopted a more conciliatory approach. These shifts continued during the War and as Beardsley points out, the sharing of secrets with the Soviet Union was "a matter largely of balancing advantage against risk".[64] Contextualizing the indecision was, of course, the ideological differences between the democratic and capitalist West and the totalitarian, communist USSR—differences that would culminate in the Cold War. As a medical instrument, the Brenthurst Splint was neither essential nor strategic in the war effort, yet Penn articulated the suspicion held by the West that was compounded by the USSR's lack of transparency. In his autobiography, he discusses the many splints sent to the USSR when the latter realized its existence. Used to treat Soviet soldiers, it was subsequently claimed as a Soviet invention, the work of "a man with an unpronounceable name". Penn also maintained a professional relationship with the head of the Moscow Traumatological Institute which he subsequently discontinued as the Russian was either oblivious to the progress made in reconstructive surgery "or he wasn't telling".[65] For Penn then, the Soviets were either behind or they did not adhere to the sharing of

[64] E.H. Beardsley, "Secrets between Friends: Applied Science Exchange between the Western Allies and the Soviet Union during World War II", *Social Studies of Science*, Vol 7, No 4, 447–473, 1977.

[65] Penn, *The Right to Look Human*, p83.

information that marked the relationship between the Allies. Either was anathema.

In contrast to the international distribution of the splint that was usually accompanied by a flurry of correspondence between Penn and the UDF, the domestic distribution of the splint was a relatively calmer affair—although the demand was as great. From as early as August 1942—a year prior to its actual distribution—interest in the splint was expressed by local medical practitioners. Writing to Watt, the Medical Superintendent of the Government Hospital in Durban requested the procedure to order the splint as soon as it was available. The Hospital was both a civilian and a military one and Durban's strategic location as a port city and stopping-off point for troopships, meant that he expected to be inundated with casualties in the event of a "blitz".[66] A request was even received from Norman Petersen who—as discussed in the previous chapter—was the other surgeon who accompanied Penn for their early training under the Big Four in Britain. Petersen had subsequently been stationed at Voortrekkerhoogte, the Springfield Hospital in Durban and in Egypt. Once discharged from military service, he continued his private practice, specializing in reconstructive surgery. With the many civilian cases of fractured jaws and the existing "inferior methods" of treatment, he felt the splint would be a marked improvement.[67] Soon after Penn's request for "gifts" to be made to the branches of empire's various military medical services, a representative of the Transvaal Dental Association wrote of the "essential" value of the splint in the civilian sphere once the military needs had been met.[68] Taking Prentice's letter into consideration and in preparation for future requests of such a nature, Lt. Colonel Watt of the DGMS used the opportunity to emphasize that the splint was only available through the UDF, its distribution should be handled by its Surgical Supply Houses and that 10% should be added to the existing

[66] Medical Superintendent, Government Hospital, Durban, 20 August 1942, File MD 540/6 MVS 615/42: "Brenthurst Splints", GG, Group 2, Box 1101, DOD.

[67] Norman Petersen, 4 May 1944, File MD 540/6 MVS 615/42: "Brenthurst Splints", GG, Group 2, Box 1101, DOD.

[68] M. Prentice, Transvaal Dental Association, 17 December 1943, File MD. 540/6 MVS 615/42: "Brenthurst Splints", GG, Group 2, Box 1101, DOD.

cost of producing the splint.[69] As the splint was the property of the UDF, Penn was not consulted.

In 1945, as the work of the Inventions Committee drew to a close, a meeting by the committee on 6th June recognized the approval of £50 each awarded to Sergeant Schulmeister and Sergeant Roylance for their role in the development of the acrylic eye and the Brenthurst Splint.[70] Penn, who had had his initial request to have the inventors at Brenthurst be sent to the United States to demonstrate their work denied, planned his own trip overseas. With his war service coming to an end, Penn requested two boxes to use in demonstrations on his visit to the various maxillo-facial centres in Britain and the United States. If the UDF did not see fit to give him the boxes, he requested them as a loan.[71] W. Grant Mackenzie wrote to the UDF, describing both the evolution of the splint and Penn's experience in the United States where on arrival, Penn was chagrined to discover that the splints intended as gifts for the institutions and leading figures in reconstructive surgery three years earlier had as yet not been dispatched from South Africa—and it was now a year after the War had concluded. To pour salt on the wound, Penn was informed that the "complications" arising from the older and invasive "pin method" of splintage had proved so severe that medical professionals were forbidden from using the method on American military personnel. This was in contrast to the relative absence of complications arising from the use of the splint at the Brenthurst Military Hospital. Yet Penn was given a warm welcome by his American colleagues and introduced to their methods of splintage. He was further asked to speak about the splint and other South African innovations in reconstructive surgery at the conference of the American Association of Plastic Surgeons held in Toronto, Canada.[72] As a civilian at the University of the Witwatersrand Dental School, Penn would later castigate the UDF for its failure to adequately encourage the production and dissemination of the splint; once its utility

[69] J.M. Watt, "Brenthurst Splint", 28 December 1943, File MD 540/6 MVS 615/42: "Brenthurst Splints", GG, Group 2, Box 1101, DOD.

[70] Inventions Sub-Committee Meeting Minutes, 6 June 1945, File 320: Minutes of Weekly Meetings of Inventions Sub-Committee, KMG, Group 27, Box 178, DOD.

[71] Jack Penn, "Brenthurst Clamp Splints", 11 January 1946, File MD 540/6 MVS 615/42: "Brenthurst Splints", GG, Group 2, Box 1101, DOD.

[72] W. Grant Mackenzie, 11 September 1946, File MD 540/6 MVS 615/42: "Brenthurst Splints", GG, Group 2, Box 1101, DOD.

was evident, demand for it could only grow and the UDF was therefore "missing a great opportunity".[73]

On completion of his tour to the United States, Penn wrote to Brigadier Du Plessis at the DGMS asking if he would be permitted to retain the splint as he was "badly in need of one and cannot possibly afford to buy one at the price named". He concluded his missive with the view that his request was "fair" as he had been accorded "no recognition…for the time, personal expense and labour" that had characterized the development of the splint.[74] The barb struck home and the DGMS authorized two complimentary sets of splints to be given to Penn and Brown as an acknowledgement of their role in this South African invention. Penn was asked to consider whether he would be willing to consider the splint already in his possession as a "gift from the Defence Department".[75]

The demand for the Brenthurst Splint, heightened by war, would decline significantly after 1945. By 1946, forty-four sets of splints had been sold, both in South Africa and overseas (including the sixteen that eventually found their way to the Soviet Union). A further fifty-two sets had been donated to the Directorate of Medical Services in the Middle East and to the British and the Americans. And a final thirty-two were still held by the UDF, whether in stores or sent to medical units; two of these were held for its inventors, Jack Penn and Lester Brown. The thirteen remaining splints were considered "surplus to military requirements".[76] They could subsequently be found at "surplus stores" and an ambitious entrepreneur acquired them for a fraction of their original cost. He, in turn, sold them to Penn for £100 per box.[77] They had come full circle.

[73] Jack Penn, 7 January 1947, File MD 540/6 MVS 615/42: "Brenthurst Splints", GG, Group 2, Box 1101, DOD.

[74] Jack Penn, 13 September 1946, File MD 540/6 MVS 615/42: "Brenthurst Splints", GG, Group 2, Box 1101, DOD.

[75] Major P.N. Brink, DGMS, "Brenthurst Clamp Splints. Application for Financial Authority SEC. 314/46", 15 November 1946, File MD 540/6 MVS 615/42: "Brenthurst Splints", GG, Group 2, Box 1101, DOD. The wording of the letter mirrored that of Penn's previous missive reading, "…in recognition of the time, personal expense and labour which they sacrificed…".

[76] Captain W.A.C. Davidson, SSO Central Medical and Veterinary Stores, "Brenthurst Splints", 20 November 1946, File MD 540/6 MVS 615/42: "Brenthurst Splints", GG, Group 2, Box 1101, DOD.

[77] Penn, *The Right to Look Human*, p83.

The Brenthurst Splint is unique in giving insight into the relationship between Allies and technology. As it was not a weapon, the UDF and the reconstructive surgeons involved in its creation had little qualm in advocating its distribution as a humanitarian gesture. Yet, because Penn and Brown had waived their rights to it, the splint was the property of the military which sought to safeguard its financial investment while simultaneously asserting its origin as a purely South African invention that was an improvement on the devices originating from the great centres of reconstructive surgery in Britain and the United States. This was in contrast to the more deferent role taken by the UDF with the invention of the flail tank as discussed in Chapter Four. Paradoxically however, the UDF did little to ensure the efficient publicization, production and dissemination of the splint. For one of its inventors, Jack Penn however, the possible financial benefits of the splint evoked no interest when measured against its widespread dissemination and use. The last not only promised greater "succour" but it was a way to demonstrate the innovative work of Penn and his colleagues in the Brenthurst Military Hospital and the skill that could be found in a Dominion country distant from the main theatres of war.

While the Union of South Africa was a relatively minor player in the Allied effort of the Second World War, for reconstructive surgeon and Office-in-Charge of the Brenthurst Red Cross Military Hospital for Plastic Surgery, Jack Penn, the inventions and innovation emanating from the hospital were used to showcase South African technological and medical achievements inspired by the War. Penn had strong views on what constituted innovation and this related to what he perceived to be "genius", a happy accident of genes. It allowed for the flourishing of the "nonconformist", a flourishing however that was somewhat restrained by what he described as the "conservative mediocrity" of the senior officers of the UDF.[78] Penn's view was shaped in part by his encounters with military bureaucracy as evident in the dissemination of the Brenthurst Splint, with the UDF privileging the pragmatic and the pecuniary over the aspirations of the inventor. As such, the inventions in reconstructive surgery illustrate an Allied relationship where the sharing of technology was marked often contradictorily by proprietorship, collaboration and competition.

[78] Penn, *The Right to Look Human*, p74 and Jack Penn, *Reflections on Life* (Johannesburg, Ernest Stanton Publishers, 1980), p33.

CHAPTER 7

Man and Machine: Ernst Jokl, Aviation Medicine and "a Rational Science of Organism"

Over the sand hills of the North Carolina coast yesterday, near Kitty Hawk, two Ohio men proved that they could soar through the air in a flying machine of their own construction, with the power to steer and speed it at will.[1]

On a windy day in North Carolina on 17 December 1903, the Wright Flyer took off on a flight lasting just twelve seconds. It was the first "heavier than air" flight and the pilot was Orville Wright who lay prone on the aircraft on the lower wing, a position that would enhance the "instinctive sense" of flying and mark the harmony of man and machine. The Flyer itself had a wingspan of just over twelve metres and weighed 274 kilograms. It was a biplane, its wings made of "wooden ribs" covered in muslin. Piano wire was used to manipulate the wings and the plane

[1] Harry P. Moore and Keville Glennan, *Virginan-Pilot*, 18 December 1903, reproduced in Chiles T.A. Larson, "The Big Story", *Virginia Living*, https://virginialiving.com/culture/the-big-story/, accessed 11 January 2024.

Attributed to L.J. Henderson and quoted in Ernst Jokl, *Aviation Medicine* (Cape Town, Unie-Volkspers Beperk, 1942), pv.

S. Chetty, *Science and Technology in South Africa, 1939-1946*, Palgrave Studies in the History of Science and Technology,
https://doi.org/10.1007/978-3-032-05084-7_7

was powered by a 12-horsepower gasoline engine.[2] Based on the Wrights' observations of birds in flight, the wings of the biplane could be "warped" or "twisted" so as to allow for the plane to change direction. The plane was also equipped with a rudder and the pilot—in his prone position—could move both the wings and the rudder through the "movement of his hips", unsurprisingly similar to riding a bicycle.[3]

Forty-four years later in the skies above the Mojave Desert in California, the Bell X-1 piloted by Chuck Yeager, broke the sound barrier. The X-1 had been carried in the bomb bay of a B-29 and was released at six thousand metres. The X-1 had a wingspan of 8.5 metres, weighed more than 5 and a half thousand kilograms, was largely composed of aluminium with a rocket engine giving it a speed of 1,556 kilometres per hour. At the time of Yeager's flight, little was known of the repercussions for man and machine for breaking the sound barrier. Earlier fatal crashes suggested the incompatibility of flight with the "invisible barrier" with concerns over turbulence and the lack of response of controls. This was compounded by the G-forces as a result of acceleration and the effects of altitude. In such challenging conditions it was the "temperament" of Captain Charles Yeager that apparently gave him the ability to "push the outside of the envelope". Six years later, Jacqueline Cochran became the first female aviator to breach the sound barrier in an F-86 Sabre equipped with a jet engine.[4]

The dramatic developments in aviation from the Wright Flyer to the Bell X-1 also required account to be taken of the requirements of the human pilots and the capabilities of their bodies. According to David Edgerton, the feature that distinguished the inventions of the Wright Brothers from the gliders and the hot air balloons that preceded them was that: "they sought human mastery of the air rather than its mastery by machine…[their machines] were sensitively controllable extensions of the human pilot".[5] The machine was subordinate to the will of the pilot—yet the body of the pilot also depended on the machine in an environment

[2] R.G. Grant, *Flight: The Complete History of Aviation* (London, Dorling Kindersley Limited, 2017), pp26–29.

[3] David McCullough, *The Wright Brothers* (London, Simon and Schuster, 2016), pp38–39, 79. The Wright brothers had developed their flier in their bicycle workshop.

[4] Grant, *Flight*, pp266–271.

[5] David Edgerton, *England and the Aeroplane: Militarism, Modernity and Machines* (Penguin, 2013), p3.

that was antithetical to the sustaining of human life. Aviation medicine is thus necessarily intertwined with the development of the aeroplane—its ideological and militaristic origins, its evolving technology and its human operators.

Despite rather paltry air strength in September 1939, South African leaders were cognizant of the importance of aircraft in the decades preceding the Second World War. It was no less a figure than Jan Smuts who, in July 1917, was accredited with the Smuts Report that made the recommendation for a branch of the armed forces distinct from the Army and the Navy. This "separate air [service]" would become the Royal Air Force. As it stood, the existing system subordinated air power to the more traditional armed services yet, for Smuts, air power had the potential to change the nature of warfare and bring it to the home front: "And the day may not be far off when serial operations with their devastation of enemy lands and destruction of industrial and populous centres on a vast scale may become the principal operations of war…" Air power was a modern form of weaponry to be used in a new kind of warfare where "Man-power…will more and more tend to become subsidiary and auxiliary to the full development and use of mechanical power".[6]

As Tilman Dedering shows, Smuts's enthusiasm for and perception of the possibilities of air power were felt in his home country. An attempt to recruit South African men "of pure European descent" for the Royal Flying Corps in the latter years of the First World War however, demonstrated that the aeroplane was very much associated with imperial power

[6] "War Cabinet Committee on Air Organisation and Home Defence Against Air Raids (2nd Report)", 17 August 1917, CAB 24/22/58, National Archives, Kew. In Tilman Dedering's discussion of the Smuts Report, he draws parallels with Smuts' independent realization of the importance of aerial warfare which mirrored that of Giulio Douhet considered the pioneering advocate for the utility of air power in warfare. Dedering further locates Smuts' views on air power within both the military and the ideological context of the period in the wake of Zeppelin raids on London. However, despite the acclaim given to the Smuts Report as the prescient embodiment of the strategic importance of aerial warfare, its visionary status has been challenged with claims that Smuts lacked the necessary military background with one scholar suggesting South Africa's "isolation…from the global centres of technological innovation". While Dedering acknowledges Smuts' "amateur" status, he nevertheless addresses the context in which the Smuts Report was produced—the fears aroused by air raids and the need for a counter in a new form of warfare. Please refer to Tilman Dedering, "'Prophet of Air Power': Jan Smuts and the History of Aviation", Michael Eckhardt (ed.), *Mission Afrika. Geschichtsschreibung über Grenzen hinweg. Festschrift für Ulrich van der Heyden* (Stuttgart, Franz Steiner, 2019), pp401–416.

and contextualized by the enmity of Afrikaner nationalists. Smuts also requested that the recruiter, Major Allister Macintosh Miller, undertake "demonstrations flights" to demonstrate air power to potentially subversive "natives" and nationalists; aircraft was from the outset perceived to be a tool of intimidation and the enactment of state power. This vision was rendered concrete when air strikes were used against striking white mineworkers in 1922 in Johannesburg and during the Bondelswarts Rebellion that same year in South Africa-controlled South West Africa.[7] Yet, even in the civilian sphere, aviation—like other forms of technology—was not immune from taking on an ideological bent. Oswald Pirow was Minister of Defence during the interwar period. Later to acquire notoriety as a Nazi sympathizer with his establishment of the right-wing New Order, Pirow perceived civilian aviation in the form of South African Airways as a means of entrenching South African hegemony in Africa while challenging British paramountcy.[8] And as Dedering argues, the trans-African flight of the *Silver Queen* in 1920 piloted by Afrikaners Quintin Brand and Pierre van Ryneveld (who would later be instrumental in the formation of the South African Air Force), symbolized both the incorporation of Afrikaners into the narrative of modernization while simultaneously evoking both a sense of awe and of intimidation in indigenous Africans, thereby excluding them from that same narrative.[9] As such, the aeroplane in South Africa symbolized—in line with Edgerton's view—militarism and the exertion of power, whether based on class, race, nationalism or South Africa's own imperial ambitions.

In *England and the Aeroplane*, Edgerton dismisses the overwhelming perception of aircraft as a civilian technology, one that would connect the world and represent a utopian harmonizing vision of modern technology. In contrast, he argues that from its very inception, the aeroplane was considered a potential weapon of war and that has since been its primary

[7] Refer to Tilman Dedering, "Air Power in South Africa, 1914–1939", *Journal of Southern African Studies*, Vol 41, No 3, Special Issue: South African Empire, June 2015, pp451–465.

[8] Robert L. McCormack, "Man with a Mission: Oswald Pirow and South African Airways, 1933–1939", *Journal of African History*, 543–557, 20, 1979, pp543–545.

[9] Refer to Tilman Dedering, "Blue Skies into White Space: Southern African Responses to the Trans-African Flight of the *Silver Queen*, 1920", *Technology and Culture*, Vol 59, 289–312, April 2018.

role.[10] Simultaneously, the portrayal of female pilots in the popular consciousness during the interwar period was marked by two competing and antithetical perceptions. The first was the use of female aviators to domesticate air travel, to suggest its safety and suitability for families. On the other hand, prevailing views of the idealized role of women in the domestic sphere and the concomitant belief in the limitations of the female body suggested their biological unsuitability as pilots.[11] While Edgerton does not specifically address the gendered nature of aviation, this chapter suggests that the contradictory portrayal of women in aviation stems from this dichotomy of the civilian and the military—the first could countenance the presence of female pilots; the second did not—and it is largely the militaristic use of the aeroplane that dominated in the first half of the twentieth century. In addition, this chapter draws upon the comprehensive history of aviation medicine by Douglas Robinson as well as works dealing in particular with trends in German and American aviation medicine. The main area of analysis here is the work of Ernst Jokl that demonstrates both the militaristic and masculine emphases in aviation medicine.

Ernst Jokl was born in Germany in 1907. He studied at the University of Breslau where he would eventually become the Director of the Institute of Sport Medicine in 1931. The coming to power of the Nazis in 1933, coupled with their virulent anti-Semitism, meant that Jokl was relieved of his position and subsequently left for South Africa. His medical degree was, however, not recognized in South Africa and Jokl had to obtain a medical degree from the University of the Witwatersrand in 1936 that would permit him to practice medicine in the country.[12] From as early as 1933 however, Jokl was associated with the Medical Corps in the Union Defence Force (UDF), as part of "a rehabilitation experiment" contextualized by the "poor white problem". This referred to the extreme poverty—with its social and physical effects—as experienced by whites which was the subject of a five-year study funded largely by the Carnegie Corporation. Underlying this was the fear of degeneration and, in a report to Oswald Pirow, Jokl stated that the "deterioration

[10] Edgerton, *England and the Aeroplane*, ppxxi–xxii.

[11] Refer to Susan Butler, *East to the Dawn: The Life of Amelia Earhart* (Cambridge, Da Capo Press, 1999).

[12] Anell Stacey Daries, "The History of Physical Education at Stellenbosch University" (Stellenbosch University, Unpublished PhD Dissertation, 2023), Chapter 2.

of individuals or groups [as a result of environmental conditions]" was not hereditary and they could thus be "rehabilitated". Young white men from poverty-stricken backgrounds were subsequently recruited into a "special battalion" at the Roberts Heights military base in Pretoria, with physical training and "military discipline" used as a means of rehabilitation. Jokl co-authored the resulting publication, *Training and Efficiency: An Experiment in Physical and Economic Rehabilitation* with its cover depicting a male figure, dressed haphazardly with poor posture and his transformation into a uniformed soldier standing stiffly at attention.[13] Another co-author was Eustace Henry Cluver, head of the South African Institute for Medical Research. The findings demonstrated the transformation wrought by physical exercise, recreation and diet of the young men of the Special Service Battalion: "They no longer gave the impression of being 'dull'. They were dressed in well-fitting uniforms which contributed to their looking 'smart'. They understood questions and learned how to answer when spoken to....A widespread and serious social disease had been cured".[14] At Roberts Heights, Jokl also worked alongside Danie Craven—a significant figure in South African rugby—and the two men produced a report on their observations regarding the physiological effects of physical exercise on the recruits, concluding that exercise "tends to establish...[the] ideal body pattern".[15]It was in his capacity as a consultant to the Director General of Medical Services in the UDF that Jokl addressed aviation medicine. His work would incorporate his experience in physical education and the "ideal" male body with the physical requirements for flight.

Jokl's Treatise

It is self-evident to state that aeroplanes are a product of the twentieth century and hence uniquely modern in a manner that much other military technology is not. As such, in aerial warfare, strategists were starting

[13] Ernst Jokl, *South African Reminiscences* (Ernst Jokl, 1988), ppxi, 22–25, 29. It should also be noted that Jokl's research would also focus on the health, nutrition and physical fitness of African, Indian and Coloured children—Refer to Jokl, *South African Reminiscences*, p26. This is an historical designation of racial classification in South Africa.

[14] "S. Africans have High Physical Potential", *Rand Dail Mail*, 15 July 1941.

[15] Ernst Jokl, *South African Reminiscences* (Ernst Jokl, 1988), p29.

anew; they had no lessons from the past to guide them and it was experience that would prove their training ground.[16] And the same can be said for the evolution of aviation medicine which had to keep pace with the changes in aerial technology.

At the outset of the First World War—and a little more than a decade after the historical flight at Kitty Hawk—there was a cavalier attitude towards the safety of pilots. A still new technology, aircraft were unpredictable, and the ideal pilot was a risk-taker, smitten by adventure and danger—and flying offered both. Parachutes were rare, physical examinations were rudimentary and training even more so. In 1914, out of every hundred British aviators killed, only two were due to enemy action; the remainder were due to the lack of health and safety standards. This would be rectified with the formation of the Royal Flying Corps. With its entry into the War in 1917, the United States would soon follow suit, establishing a Medical Research Board to determine the factors responsible for the "efficiency" of aviators; their tolerance for high altitude and the provision of oxygen for such; and to test their fitness levels. These would become the core features of aviation medicine.[17] Its context was the First World War, and an American contingent was dispatched to the Western Front to observe "the conditions under which pilots lived and flew their planes". This revealed both the hierarchy and a certain sense of bravado and bravery where pilots required authorization from a superior officer to seek medical attention, with many opting not to do so for fear of being seen "as lacking in courage".[18] German aviation medicine also had its origins in the First World War but was halted by the Treaty of Versailles that imposed restrictions on aviation. In 1927, however, August Ludolph Brauer would establish the German Institute for Aviation Medicine with an initial focus on the physiological effects of high altitude. The Institute was home to pioneering figures in aviation medicine in the 1930s such as

[16] Paul Kennedy, *Engineers of Victory: The Problem Solvers Who Turned the Tide in the Second World War (*Penguin 2014), pp81–82.

[17] Green Peyton, *Fifty Years of Aerospace Medicine: Its Evolution since the Founding of the United States Air Force School of Aerospace Medicine in January 1918* (AFSC Historical Publications Series No. 67–180, 1968), pp2–3, 8.

[18] Jean Evans, "History of the School of Aviation Medicine, 1918–1958", *The Air Power Historian,* Vol 5, No 4, 245–261, October 1958, p247.

Heinz von Diringshofen[19] whose work would be incorporated in Jokl's *Aviation Medicine*.

In an article appearing in a South African newspaper seven months before the outbreak of war, mention was made of the arrival of Hawker Hurricane fighters in South Africa that would be used to bolster the country's air power. These aircraft were to acquaint Air Force pilots with the increasing speed and manoeuvrability associated with the "modern fighter" and prepare them for active service. It was in line with comparable shipments to Australia and New Zealand and similar aircraft were also envisioned as protecting Britain. Accompanying the article on the aircraft was an excerpt on aviation medicine with Jokl's advocacy for the importance of the field and the obtaining of "experts" and knowledge that would enable South Africa to "stand on its own two feet" in the event of conflict.[20] In South Africa then, aircraft and aviation medicine were intertwined, both related to a sense of empire as well as national self-sufficiency—and both would ultimately be focused on the male body.

For Jokl, the scientific principles that applied to aircraft design could resonate in the human body: a plane's performance could be predicted scientifically as could the accuracy of its weaponry and so, too, could predictions be made regarding "a pilot's reactions to the strains and shocks of aerial combat". It was the conjunction of scientific knowledge about flight and the human body that formed the core of aviation medicine.[21] Further, aviation medicine was not only based on a combination of scientific thinking and practice—for which war gave ample opportunity—but Jokl believed it to be collaborative. He singled Germany out for its early advances in aviation medicine, particularly in "dive-bombing and mass parachute attacks".[22] A pioneer in aviation medicine was Nathan Zuntz. Like Jokl, Zuntz was of German-Jewish extraction and graduated with his medical degree from the University of Bonn in 1868. His dissertation focused on the physiology of blood and, from relatively early in his career, his research addressed the effects of pressure and oxygen content on respiration and would later include metabolism as well as the

[19] Viktor Harsch, "Ludolph Brauer: German Aeromedical Pioneer", *Aviation, Space and Environmental Medicine,* Vol 5, No 8, August 2004, pp705–707.

[20] "300 m.p.h Fighter Planes for Union" and "'Aviation Medicine' Urged", *Rand Daily Mail*, 3 February 1939.

[21] Jokl, *Aviation Medicine*, pv.

[22] Jokl, *Aviation Medicine*, pvi.

digestive process. This would lead him to consider the ways in which the body behaved under "strain", with his aim to both delineate the limitations of human endurance and to enhance "performance". While much of his experiments used animals, Zuntz was not averse to being a test subject himself and, along with his research partner Adolph Loewy considered the ways in which the human metabolism was affected by a drop in oxygen levels—as would be experienced at high altitude. His findings were expressed in *Studien zu einer Physiologie des Marsches* (Studies on the Physiology of Marching) published in the late nineteenth century that was part of a move by the German War Department to apply medical science to render the army a more effective fighting force.[23] Science and medicine were harnessed to both the understanding and improvement of the human body and with figures such as Zuntz, Germany was at the forefront of what would become aviation medicine.

However, by the mid-point of the Second World War a new phase was beginning, one that would steadily turn the tide in favour of the Allies. Simultaneously, from the rise to power of the Nazis in 1933, Germany had prioritized ideology over science, marginalizing and expelling Jewish scientists—including Jokl—who were replaced by their preferred race counterparts that may have lacked experience and/or ability. With the outbreak of War, Germany was isolated from the international channels that fostered scientific learning and exchange, eventually forced to use illegal means to obtain scientific journals.[24] Jokl describes German's subordination of science to ideology as a "self-imposed intellectual blackout" that was in contrast to the collaborative effort among the Allies. Developments in American aviation—the technological superiority of their bombers—also meant that the Allies were taking the lead in aviation medicine.[25] Yet, at the time of Jokl's writing, Germany was continuing its research into aviation medicine. In the interwar period, Hubertus Strughold had carried out physiological experiments on himself to assess the effects of flight, considering the limitations on co-ordination and cognition due to lack of oxygen as well the effects of altitude on vision. In 1935 Strughold was the head of the Aviation Medical Research

[23] Hanns-Christian Gunga, *Nathan Luntz: His Life and Work in the Fields of High Altitude Physiology and Aviation Medicine* (Elsevier Inc, 2009), pp7, 14, 21, 26–27, 30.

[24] Tom Shachtman, *Laboratory Warriors: How Allied Science and Technology Tipped the Balance in World War II* (New York, Perennial, 2003), pp41, 100.

[25] Jokl, *Aviation Medicine*, pvi–vii.

Institute in Berlin and no expense was spared in terms of the experiments carried out using a low-pressure chamber and a centrifuge. With the advancement in aviation technology—Germany was the first to effectively employ the jet engine in the Messerschmitt Me 262—as well as new theatres of war, additional "Experimental Stations" were established to test the new conditions of flight. Further experiments would be conducted on concentration camp inmates, studying the effects of altitude and thirst (for those pilots who had made forced landings at sea), as well as the effects of freezing on human physiology. These unwilling subjects often succumbed to these experiments. Strughold and his collaborators in aviation research were recruited by the Americans after the War as the age of supersonic flight began.[26]

With *Aviation Medicine*, Jokl made his own contribution. He considered his book to be unique, the first of its kind with the only comparison being *Fit to Fly* by Grow and Armstrong of the United States Air Corps. However, while their handbook on aviation medicine addressed the more "general aspects of health and fitness for aviation", Jokl was convinced his work focused on more "specific medical problems" likely to arise from the conditions associated with flying: "altitude, blind flying, speed and parachuting".[27] The text is, however, largely a compilation of the existing knowledge regarding aviation medicine in the United States, Britain and in Germany—with some input made by Jokl regarding his own experiences as physical education instructor and as a pilot.

Even prior to the historical flight by Orville Wright in 1903, there was an awareness of the effects of altitude on human physiology. Hot air ballooning had been taking place since the late eighteenth century, and it was only prior to the flight by the Montgolfier brothers that oxygen had been identified as a gas by Joseph Priestley. There was therefore little understanding of the effects of lack of oxygen on the human body at higher altitudes that would subsequently be termed anoxia. Almost simultaneously with the invention of the hot air balloon, was the growing European passion for mountaineering.[28] Until the eighteenth

[26] Annie Jacobsen, *Operation Paperclip: The Secret Intelligence Program that Brought Nazi Scientists to America* (New York, Back Bay Books, 2015), pp209, 211–213, 215, 217.

[27] Jokl, *Aviation Medicine*, pvii.

[28] Douglas H. Robinson, *The Dangerous Sky: A History of Aviation Medicine* (Henley-on-Thams, GT Foulis and Co Ltd, 1973), pp1, 9–10.

century, mountains were forbidding and inaccessible with those forced to contend with them giving descriptive accounts of the effects of altitude—which they sought to counter by holding "vinegar-soaked sponges to their mouths", to no avail. With the onset of mountaineering among Europeans, the physical toll of ascending ever higher peaks coupled with altitude sickness, made the pastime an expression of masculinity that was associated with the sublime—the "dazzling brilliancy" of the natural world and the successful pitting of will and strength against unyielding rock.[29] Yet success was not easily achieved—if at all. The symptoms of anoxia ranged from the relatively mild—an adverse effect on night vision at relatively low altitudes—to increased impairment at higher altitudes: "fatigue…dizziness…an unwarranted sense of well-being resembling alcohol intoxication", poor cognitive skills and "muscle coordination". With increased altitude, a person will fall unconscious and should oxygen not be administered, death ensues.[30] All of these were addressed in painstaking detail in Jokl's chapter, "The High Altitude Problem".

Jokl's focus was on the experience of mountaineers that had provided the foundations for an understanding of altitude sickness with one important caveat: climbing a mountain was gradual whereas an aeroplane could potentially ascend at a breath-taking nine hundred metres a minute. A solution for looking at the effects of altitude then could therefore be found in the "easily controllable environment of the laboratory". Jokl went on to describe the use of early pressure chambers by French scientists in the nineteenth century. This was subsequently adopted by Jokl's teacher, Adolf Loewy in Germany who was able to carry out a detailed analysis of the effects of low pressure arising from altitude.[31] Loewy and Nathan Zuntz made use of a "pneumatic chamber" at the Jewish Hospital in Berlin that was used as part of the therapeutic process, exposing patients to incremental adjustments in air pressure. The controlled conditions of the chamber allowed the scientists to supplement their earlier research in the field where they ascended the Alps and more modestly

[29] Robert MacFarlane, *Mountains of the Mind: A History of a Fascination* (London, Granta Books, 2008), pp145, 162–163.

[30] Robinson, *The Dangerous Sky*, pp10–11.

[31] Jokl, *Aviation Medicine*, pp22–24.

carried out "climbing experiments" on a staircase at the College of Agriculture.[32]

Using the work of his German forebears and others, Jokl compiled a detailed account of the effects of altitude (and lack of oxygen) on pilots that included "acute mental failure" with an accompanying case study of an American pilot suffering from anoxia due to a malfunctioning altimeter which placed him at a lower altitude than he actually was. The pilot found it difficult to use the microphone to radio in and did not even contemplate the use of oxygen. Despite his difficulty, he nevertheless continued his ascent. It was an illustration of impaired judgement. Physiological effects included the expanse of bodily gases and the painful effect on the sinuses and the ear should the pilot have a cold. Solutions to the problem of altitude included a pressurized cabin, the use of "pressure suits" akin to those worn by deep-sea divers, oxygen masks and tanks.[33] Underlying Jokl's analysis was the context of war where "the conquest of the stratosphere" was presented as a "military problem". Jokl described the efficacy of bombers such as the Flying Fortress flying at high altitude that would place it out of the range of fighter aircraft. He dismissed the pressurized cabin system as a solution for high flying aircraft, drawing attention to the dangers of the cabin being pierced by enemy fire. The preferred solution was the use of oxygen administered via facial masks. Test subjects were those who had already demonstrated the ability to withstand the effect of low pressure as administered in low pressure chamber tests. These chambers however were few and far between while the War necessitated an increase in training. The solution was simply the administering of a lesser concentration of oxygen to mimic what the pilot would be exposed to at altitude. While convenient, the drawback was that it did not permit the study of actual low pressure on the body. Tests also demonstrated that the sudden availability of oxygen to a person deprived of it had detrimental effects creating a "shock" to the system and resulting in loss of consciousness. Jokl drew upon his own experience in physical education to illustrate

[32] Gunga, *Nathan Luntz,* pp19, 27–28.

[33] Pilots during the war were fitted with "G suits" where the pants contained "air bladders". With the increase in G-forces, the air bladders would constrict blood flow to the lower half of the body, ensuring that it remained above the waist and thus preventing loss of consciousness—Refer to James Bradley, *Flyboys: A True Story of Courage* (New York, Back Bay Books, 2004), p129.

this describing the example of sprinting athletes who, upon the conclusion of the race, often lost consciousness even if they were "outstanding men".[34]

To overcome the worst effects of altitude, pilots had to in fact be "outstanding men". An important feature was the health of the adrenal ("suprarenal") glands, considered to be important for fighting the effects of fatigue. Pilots had to be fully recovered from an "infectious illness", including a minor cold—as poor health is worsened under flight conditions. And, again, related to Jokl's area of expertise, physical training was necessary to maintain the fitness of the pilot. The training was however not to be too strenuous as flying was already a "strain" on the body or the "organism". In addition, the playing of games such as "water play" and "easy exercise"—if they did not demand too much in the way of exertion and were not made compulsory—would ease potential mental fatigue as well.[35]

While addressing the effects of altitude became an ever greater concern as aircraft became more sophisticated, another feature of aircraft that changed dramatically was speed as engines became more powerful and gasoline engines and propellers gave way to the jet engine—and, during the Space Race—the rocket. As Jokl pointed out however, it was less about speed itself than about abrupt changes in speed and direction, confronting the "organism" with centrifugal force—or G-force.[36] Due to the speed and manoeuvrability of military aircraft, the effects of G-forces were usually studied under military rather than civilian conditions and were provoked by acceleration as well as turns. Pilots also experienced high G-forces when ejecting—although this was for a short period with little repercussion. The effects of positive G-forces were dramatic. At three Gs, the body weighs three times its original weight, making it impossible to walk. The internal organs are temporarily forced downward and blood flows away from the brain, amassing in the abdomen and legs. This may result in decreased vision followed by loss of vision and unconsciousness. This is however temporary, and no lasting effects occur.[37]

[34] Jokl, *Aviation Medicine,* pp29–30, 35, 37, 44–45, 50, 52, 63–64, 80.

[35] Jokl, *Aviation Medicine,* pp82–83, 85.

[36] Jokl, *Aviation Medicine,* p101.

[37] Robinson, *The Dangerous Sky*, p140.

Again, as Jokl shows, it was the Germans who had made the most extensive study of the effect of G-forces in the interwar period with a consideration of the likely use of dive-bombers in war. (Jokl was particularly critical of the Allies ignoring the repercussions of German research which demonstrated their dive-bombing strategy). To illustrate the effects of G-forces, Jokl drew upon the work of German pioneer, Heinz Von Diringshofen.[38] Using a biplane, Von Diringshofen—a trained pilot—began his early tests in 1931 to determine the ways in which acceleration impacted blood pressure and circulation as well as respiration. The instruments used to measure physiological changes were adapted for the plane by his engineer brother, Bernd. He would also work with Bernd in 1934 to develop a centrifuge to make a controlled study of the effects of acceleration on the human body. The results of his many experiments were published in 1939, translated into English and French and shared internationally with the RAF, among others. With the outbreak of the War however, he began the training of flight personnel while his work, *Aeromedical Guide for Aircrew*, simultaneously underpinned the training of North American flight crews destined for the RAF.[39] Also using Von Diringshofen's work, Jokl discussed the different effects of acceleration depending on the posture of the pilot who could be placed in initially both a seated and lying position, but also a standing one. Accompanying this were descriptions by the subjects of their experiences with one able to time the length of his blackout with a stopwatch and another experiencing a total loss of memory when he lost consciousness. Von Diringshofen's work had therefore demonstrated the actual "physiological realities" regarding the effects of acceleration.[40]

Jokl went on to describe in detail the animal and human experiments conducted by other German scientists related to the physiological effects of acceleration. These experiments provided insight into the ideal position for pilots to adopt to minimize the effect of G-forces. But, also evident in the studies—as was the case with altitude testing—was the varying abilities of individuals to withstand these G-forces which was known as the "collapse threshold". In addition to specially adapted cockpits and the

[38] Jokl, *Aviation Medicine,* p101.

[39] Viktor Harsch, "Heinz von Diringshofen: Aeromedical Pioneer", *Aerospace Medicine and Human Performance,* 501–503, Vol 90, No 5, May 2019, pp501–502.

[40] Jokl, *Aviation Medicine,* pp104, 107.

more modest change of position of the pilot, another potential means of dealing with acceleration was the donning of pressure suits to ensure that the blood remained evenly distributed throughout the body. Pressure suits were however at this point bulky and unwieldy, wholly unsuited for the confined space of the cockpit, and Jokl was forced to concede that, at the time of writing, there remained no definitive method of countering the effects of acceleration. In addition, special precautions were recommended for bomber crews as those in standing positions tended to suffer fractures due to the competing force of the downward exertion of gravity and the force presented by the plane altering direction. This "cross acceleration" created severe stress for the "structural stability of the skeleton" and the recommendation was that bombers (or other aircraft where men had to assume a standing position) should not exceed a force of 3 Gs. The obverse of acceleration was, of course, deceleration and fractures were evident here as well as impact injuries as was depicted in studies in the United States, drawn upon by Jokl.[41]

Deceleration, the positioning of the body and impact injuries also related to the use of parachutes. There were two envisaged uses for the parachute: the first was that it was the means by which the pilot and other aircrew could safely reach the ground in the event of an emergency. According to Robinson in his extensive history of aviation medicine, the interwar period saw the development and widespread use of parachutes—in a largely military rather than civilian context. One of the first countries to make use of the parachute towards the end of the First World War was Germany however, the United States subsequently took the lead with the "free" type of parachute that was worn in a "parachute harness" on the back and would become standardized. As early as 1922, a pilot ejecting from a bomber at an altitude of 7,300 metres carried an "oxygen flask" to prevent him from losing consciousness. He suffered airsickness while descending due to the wild "[oscillations]" of the parachute and subsequently broke the bones of his foot while landing.[42]

The other use of parachutes was less about the possible saving of lives than it was about the use of soldiers as paratroopers—and it is this offensive capability of the parachute on which Jokl focuses. On D-Day, thousands of American paratroopers were dropped behind enemy lines as

[41] Jokl, *Aviation Medicine,* pp123–125, 128, 140–141.

[42] Robinson, *A History of Aviation Medicine*, pp143–146.

part of the Allied invasion. They were the culmination of the increasing use of paratroopers, first initiated by the Soviets as a new tactical strategy as early as 1932. This also necessitated new considerations in aviation medicine. Other than the "landing injuries" as indicated in the previous paragraphs, scientists had to consider "the psychology of fear"—the unenviable sense of apprehension of having to jump out of an aeroplane at altitude—the physiological reaction to the descent, "respiratory technique during the jump" and the most effective way of placing the "strapping" of the harness. Parachutes also needed to be adapted for aircraft that flew faster and higher. An important feature was the "delayed opening" necessary first to avoid potential collisions with other parachutists. Significant however was that, if a parachute opened immediately after jumping out a fast-moving plane, it would "be torn to pieces", hence the need for a delayed opening. In terms of altitude, jumping out of a plane flying above 9,000 metres, means that the parachutist will not be able to take in sufficient oxygen for at least 3–4 minutes until he reached a safe altitude—and this could result in brain damage or death. A solution was the provision of a portable oxygen cannister that had been developed at the Mayo Clinic.[43]

To become a paratrooper then required a particularly physiological and psychological aptitude and Jokl drew upon the work done by the Americans in the form of Captain D.E. Thomas of the US Army Medical Corps: "…the avoidance of injury depends 'on a man's ability to think fast, to learn the proper technique and to make his muscles obey his mind'".[44] It was the ability to subordinate the body and the instinctive sense of fear to the rational mind. The physical requirements were stringent: a height of between 1.68 and 1.83 metres, weight of 68–84 kilograms, 20/40 uncorrected vision to determine height and objects as the jumper nears the ground, no history of previous joint injuries and "good physical strength, stamina and co-ordination [determined] by his ability to do fifteen push-ups" with a preferred "excellence in some competitive sport". In terms of mental acuity, the prospective paratrooper should be above average intelligence as determined by intelligence testing. In terms of psychology, this was still being determined during the training process and those

[43] Jokl, *Aviation Medicine*, pp175–176, 185–186, 187–188.

[44] Jokl, *Aviation Medicine*, p199.

who failed to jump would provide the basis of studying "intestinal fortitude" i.e. courage. Similar requirements were demanded of German paratroopers. Concluding his discussion on the use of parachutes, Jokl once again emphasized their military use—a civilian would feel only "anxiety" (and thus presumably lack the "intestinal fortitude") should he be fitted with a parachute.[45] To jump out of a moving plane then required physical and psychological strength with a concomitant ability to subordinate the instinctive to the cognitive. This marginalization of the instinctive and of the sensory would be evident in his discussion of blind flying.

In his chapter on "Blind Flying", Jokl uses his own experiences to consider the ways in which the individual's sensory knowledge of the position of his body in an environment to which he has not adapted (i.e. "the air") is a challenge to determine. Unlike the study of the effects of altitude which could draw upon ballooning and mountain climbing, "blind flying" related specifically to flying in aircraft and, as Jokl pointed out, the senses—sight in the absence of visual cues, ear to detect changes in pressure as well as the sensory organs of the body itself—were not acute enough to give a strong indication of the "body position in space". Simultaneously, the senses were imperfect and could potentially mislead the pilot. To illustrate, Jokl described his experience of learning how to land an aircraft. He was convinced that his plane was absolutely level, only to find that his "right wing was hanging": Neither my eyes nor my ears, nor did the combined set of peripheral sensory messages inform my brain correctly of my position in space".[46] Jokl had also learned to fly in what he termed a "primitive" environment—he used a biplane that required its wings to be "screwed…into position" and the propeller "cranked by hand". It was not equipped with instruments, relying largely on the sensory input from the pilot which, as he demonstrated, was limited.[47]

Vision was also subject to optical illusions—the appearance of movement in relation to another moving object and a sense of vertigo, bringing with it the feeling—and fear—of falling. Jokl also drew upon the findings resulting from a plane crash in East London in South Africa which highlighted the "mirror effect"—the inability of a pilot to accurately judge his height when flying over a still body of water. To illustrate the sensory

[45] Jokl, *Aviation Medicine,* pp200–202, 204.

[46] Jokl, *Aviation Medicine,* pp143–145.

[47] Refer to Jokl, *South African Reminiscences.*

fallibility of the ear in relation to balance, Jokl discussed his own work in physical education where students were subject to the Rolling Wheel. This was a large round frame within which students could stand and hold onto the frame itself. They were then "rolled". The relatively rapid change in orientation meant that the inner ear could not compensate: "After one or two turns with the wheel, they lost more or less completely their spatial orientation; they did not know when their feet were down or up, and they were unable to stop rolling when they reached the upright position". With the exception of the "mirror effect" however, the senses could eventually adapt to flight conditions.[48]

Yet the overcoming of the potential hazards associated with blind flying was largely due to the use of instruments: "the Directional Gyro, the Turn and Bank Indicator, the Rate of Climb, the Rate of Speed, and the Sensitive Altimeter". From the outset, there tended to be a reluctance among aviators to place their trust in instruments above the sensory information on which they had been reliant since birth. These were perceived to "artificial sense organs" and a poor substitute for human instinct and feeling and for Jokl represented the "fascinating and amazing competition [of] ancient biology vs. modern technique". Yet, as he noted earlier, flight was something for which the human body had not evolved and, unlike most other weapons of war, the aircraft was a truly modern machine. It therefore supplanted the human body with its sensory fallibility. Right, from the outset though, trainee pilots had to have impressed upon them the prioritization of the machine's instruments over their instinctive, sensory input. The former was a "fact", the latter a "message" from the body and "once a way [was found] of substituting objectively recording instruments for our imperfect sense organs, the battle was won. Today the battle is won".[49]

The battle was won through technology. The "Link Trainer" functioned as a kind of early flight simulator designed to accustom novice pilots to flight conditions in order to train them to learn to rely on their instruments while simultaneously demonstrating the fallibility of the senses. The inner ear was the focus of the Link Trainer for the purposes of "sensory re-education" as it was the "sensory messages" relayed by the inner ear which could result in vertigo and, in some

[48] Jokl, *Aviation Medicine,* pp145–147.

[49] Jokl, *Aviation Medicine,* pp149, 151, 153.

cases, its "hyper-sensitivity" that resulted in airsickness. It is not entirely surprising that Jokl recommended physical exercise as a solution and, in particular gymnastics, that would train the inner ear to adapt to rapidly changing orientation.[50]

The Link Trainer marked the beginning of the harmonizing between man and machine. There were four phases in the development, dating from the dawn of flight, of this harmonization which necessitated the subordination of the senses to the instruments. The "directional gyro" was an early substitute for the inner ear. Cameras could compensate for the failure of a pilot to recall the exact details of his flight and could record his actions in combat, compensating for "the inferiority of the human eye and visual memory" and replacing subjectivity with fact. Then there was radio for communication, the altimeter, and the airspeed indicator. This ultimately created "a new 'level of integration' between 'living' sense organs and 'dead' instruments" where Jokl envisioned a future where all aspects of flight were largely under the control of the machine. Through technology, the shortcomings of "a million years" of evolution could be rectified almost instantly.[51] It was the supreme expression of the machine as the embodiment of the will of the pilot as envisaged by the Wright Brothers, yet it was not clear with the increased complexity of aircraft whether it was man or the machine that was the master.

The Aviator

In his chapter on "Blind Flying", Jokl begins with a quote that pitted individual belief and physical autonomy against the necessity for medical science to be able to make broad generalizations (and hence view the individual as representative of a type), as well as the obligation of the pilot to remain in peak condition as demanded by his duty in war.[52] It was aviation medicine that connected the autonomous body of the individual to the aircraft—the modern machine of warfare—and to the nationalism of war.

[50] Jokl, *Aviation Medicine,* pp154, 156–158, 160.

[51] Jokl, *Aviation Medicine,* pp160, 162–165, 170.

[52] Jokl, *Aviation Medicine,* p143.

Yet Jokl was unclear as to how to determine the ideal aviator. To be in good health and capable of doing one task did not necessarily translate into ability to do another—and aviation demanded a great deal, both physiologically and psychologically. And a key question was how to establish a physical and a psychological norm. To address the latter, Jokl drew upon the insight provided by the United States Navy whose aviators were arguably taking part in "the most exacting branch of military flying". In addition to the US Navy, other Allies had also made use of psychologists to weed out those likely to be unsuitable for the demands of flight. Advance selection carried out in this manner however had proved unsuccessful as the tests were themselves too ambitious and resulted in a "[rejection of the] poor material [as well as] too much good material as well".[53] Psychological testing was, as yet then, inadequate and it was only through field testing that suitability could be determined: "Our proper unit for study is the man on the job". The study was envisaged as a combination of initial test data with comparisons made during the training phase and it was the student's efficacy (or lack thereof) in training that could then reinforce the validity of the test.[54]

For the physical requirements, the solution was both simple yet difficult to implement. It required the use of equipment already discussed in Jokl's book such as the Link Trainer and the low-pressure chamber however these were not always available and, if they were, often lacked experts to adequately administer, monitor and interpret the tests. A solution then to the limitations of both psychological and physiological testing was widespread training in aviation, beginning in schools and continued in tertiary institutions. Drawing upon the syllabi from secondary schools in the Soviet Union and in Germany, Jokl advocated the teaching of aviation as incorporated into subjects such as History, Physics, Chemistry and—related to aviation medicine—Biology. It was to be a modern discipline of study affiliated to the more traditional school subjects. In conjunction with this, was an emphasis on "physical education" to prepare young bodies for flight, early training in flying gliders and light aircraft and a "pre-military training programme". The very human interest in and

[53] The psychological testing of potential recruits in the South African Air Force is discussed in the following chapter.

[54] Jokl, *Aviation Medicine,* pp206–207.

passion for flying should be given a practical and "rational" outlet. And this passion—and its outlet—included "every boy and many a girl".[55]

In this essay, the anachronistic use of "he" to represent a human type has been employed, in line with Jokl's writing, and throughout his book on aviation medicine the aviator has been depicted graphically and in writing as male; while expressing his doubts regarding a norm, it is male physiology, psychology and mental acuity that is at the heart of aviation medicine yet, as he nears the conclusion, Jokl makes allowances for the female pilot when discussing the methods of selecting pilots and the study of aviation.[56] His brief concession to the involvement of women in aviation is reflective of the ambiguity regarding female pilots evident during the expansion of aviation in the interwar period.

A German veteran of the First World War described the aviator as "part of 'a new and commanding breed…a race that builds machines and trusts machines.'"[57] As Michael Paris shows, this sense of elitism associated with aviators was also evident in both the American Army Air Corps as well as the RAF, and stemmed in part from the perceived importance of air power as articulated in the Smuts Report. Buttressed by popular culture—such as Hollywood films and science fiction novels—pilots were associated with the new technology and were the brave individuals who made a remarkable contribution to the First World War, epitomizing the "one success story" in a war where much was lost with little gain.[58] Edgerton makes a convincing argument for the elitist nature of aviation that tapped into earlier concepts of honour and chivalry in warfare as evident in descriptors such as the "knights of the air" and the "cavalry of the clouds" where, in the face of the dehumanization of industrial warfare, the machine was nothing more than the equivalent of a horse, subordinate to the will of its master.[59] Flight gave aviators—as Colin Cook discusses—a sense of "control over his fate" in contrast to the infantry confined to the trenches of the First World War. With the aeroplane firmly embedded in the

[55] Jokl, *Aviation Medicine,* pp207–213,

[56] Jokl, *Aviation Medicine,* pp206, 213.

[57] Clare Mulley, *The Women Who Flew for Hitler: The True Story of Hitler's Valkyries* (London, Macmillan, 2017), Chapter 1.

[58] Michael Paris, "The Rise of the Airmen: The Origins of Air Force Elitism, c. 1890–1918", *Journal of Contemporary History,* Vol 28, No 1, 123–141, 1993, pp123–125.

[59] Edgerton, *England and the Aeroplane*, p73.

notion of the "modern", its operators perceived themselves as an elite.[60] These views presuppose the masculine (and militaristic) nature of aviation. However, civil aviation had, during the interwar period, provided scope for female aviators. While flying was uniformly an elitist occupation confined to the middle and upper classes, the exploits of Amelia Earhart were as lauded as those of Charles Lindbergh.

Despite this, female aviators were the exception rather than the rule, employed as figureheads to allay the fears of civilians and to encourage women to fly on civilian airliners. They were a novelty, evoking curiosity and their bodies believed to be unsuited as pilots. In the United States for instance, in 1942 (at the time of Jokl's writing) women were advised not to fly "three days prior to and three days after their menstrual period" and it would only be until 1945 that menstruation was determined to have "no effect on pilot performance". Prominent aviators such as Earhart, however, attempted to contend with the prevailing view of women's ultimate unsuitability and difference. While her final flight is remembered for the enigmatic disappearance of Earhart and her navigator, Fred Noonan, it was initially envisaged partly as a research flight. While testing the plane and its instruments, she was also planning to look at the effects of fatigue "on long-distance and high-altitude flights", the way in which digestion was affected by flight and which foods would be most suitable, the extent to which instruments contributed to fatigue and if there were "differences" in the ways in which men and women experienced flight.[61] It was therefore an exercise in aviation medicine for civilian aviation.

Yet, as Nazi Germany would demonstrate, there was scope for women to be involved in military aviation as well. In her compelling biography of two female pilots in Germany, Clare Mulley describes the role of pilot, Melitta Schiller von Stauffenberg, who tested dive-bombers. Von Stauffenberg would climb to 4,000 metres and then throw her Junkers into an almost vertical dive, levelling out between 150 and 200 metres above ground level. During these dives she experienced "red-out" as blood flow was diverted to her head, occasionally leading to the ["rupture of blood

[60] Colin Cook, "A Fascist Memory: Oswald Mosley and the Myth of the Airman", *European Review of History*, Vol 4, No 2, 147–162, 1997, p148. However, refer to the chapter on aptitude testing for a consideration of the particular psychological stresses believed to apply to aviators during the Second World War.

[61] Butler, *East to the Dawn*, pp226–227, 337, 359–360.

vessels in her eyes".] When she levelled out, the sudden change in direction led to high G-forces, occasionally causing her to temporarily lose her vision or even briefly consciousness. Cognizant of the perception of the weakness of the female body, she refused to concede that she was severely affected by G-forces and proclaimed her "intense concentration" to be a factor in "preventing the flow of blood away from her brain". She reputedly made up to fifteen of these punishing flights a day—far more than her male counterparts.[62] As with the "knights of the air", she exemplified the dominance of the individual over the machine and the subordination of the body to the mind.

However, even as a test pilot suffering the physiological effects of flying a bomber, von Stauffenberg was not a combatant. The situation was different in the Soviet Union where, in the interwar period women were trained as civilian pilots with some female aviators also members of the air force. With the German invasion of the motherland, the defence of the Soviet Union and the exceptional conditions of war, led to a greater acceptance of these women as combatants. This underpinned the formation of three female regiments that would serve in combat, flying both fighter planes and bombers. The 588th Night Bomber Regiment would be christened "Night Witches" by the Germans. Relegated to outdated aircraft and often dismissed by their male counterparts as "little girls", they were nevertheless formidable, engaging in night bombing missions often without the use of parachutes.[63] But female pilots in combat were the exception rather than the rule and, on a visit to the Soviet Union, Charles Lindbergh—who had no objection to civilian female aviators—was dismissive of the role of women in combat: "They command men and are commanded by men [and] are given the same missions, etc. I don't see how it can work very well".[64] It was also Lindbergh who explicitly associated particular "races" (such as the Germans) with "virility" and hence embodying idealized masculine attributes such as "courage"[65]—which were central to combat and aviation.

62 Mulley, *The Women Who Flew for Hitler*, Chapter 7.

63 Kristal L. M. Alfonso, *Femme Fatale: An Examination of the Role of Women in Combat and the Policy Implications for Future American Military Operations* (Air University Press, 2009), pp23–25, 30–31.

64 Quoted in David M. Friedman, *The Immortalists: Charles Lindbergh, Dr Alexis Carrel and their Daring Quest to Live Forever* (New York, Harper Perennial, 2008), p144.

65 Friedman, *The Immortalists*, p130.

In South Africa, the women's auxiliary arm of the Air Force was established some nine months after the country's entry into the War. The women of the Women's Auxiliary Air Force (WAAF) served as mechanics and a select few as pilots. The latter—who had qualified as civilian pilots prior to the War—also served as instructors, training potential male pilots using simulators such as the Link Trainer. Very few women actually served as pilots however—and none in a combat capacity. They were largely assigned to transporting aircraft from South Africa to the war fronts or between bases. One of these pilots, Captain Rosamund Everard-Steenkamp did become one of the early female jet pilots as an attempt to demonstrate that "even a woman could do it", mirroring the history of aviation itself.[66] As such the history of aviation in wartime South Africa was similar to that of the Allied countries—with the notable exception of the Soviet Union. Aircraft and their pilots were associated with the masculine and the militaristic. Yet, if this assumption of the male, combatant aviator is made in Jokl's book, it is more ambiguously evident in the teaching of aviation and aviation medicine.

While working as a coach at the University of the Witwatersrand, Jokl was offered a position as the inaugural head of Physical Education at Stellenbosch University.[67] In 1917–1918 as Major Miller carried out his recruitment efforts for the Royal Flying Corps in South Africa he was confronted with the "vociferous" opposition from students at Stellenbosch University (with its ties to Afrikaner nationalism) who unequivocally rejected his advances.[68] Yet it was Jokl's brief tenure at Stellenbosch in the interwar period that partly formed the basis for the development of aviation medicine in South Africa. In 1937, Jokl introduced a diploma course that would run over three years and lead to the training and development of "physical education professionals" who could engage in research and were not simply practitioners or teachers in physical education. The course reflected Jokl's interest in flight and in aviation medicine where students in their first year of study could—in addition to courses in anatomy and biology as well as engaging in

[66] Esté Mari Kotzé, "More Than Just Pretty Girls in Uniform: A Historical Study of Women's Military Roles during World War II, 1939–1945" (Stellenbosch University, Unpublished PhD Dissertation, 2021), pp131, 163, 169–170.

[67] Daries, "The History of Physical Education at Stellenbosch University", Chapter 2.

[68] Dedering, "Air Power in South Africa", pp453–454.

physical training—take "flying". Female students, unlike their male counterparts, could opt for either "flying" or a "Household course". By their final year of study, courses included Psychology, "Medical Aspects of Physical Education" and "Physics of Aviation". As Daries argues—and as Jokl's work in physical education reflects—the course was a combination of the use of scientific principles to effect social change. Jokl's course, however, did not receive approval from the Cape Education Department due to its sole focus on physical education as this would limit the employment opportunities for graduates and, especially, female graduates. In 1937 then, only those who already had a teaching degree could take the course. Jokl's tenure at the university however was too brief to assess the possibilities and impact of courses in aviation medicine on students. In a conservative environment, Jokl's physical examination of female students was contrary "to the values of the 'volk' and the university" and he was compelled to resign in April 1937.[69]

A decade later however, in a report submitted to the Union Education Department, Jokl considered the physiological differences between men and women, concluding that the "physical efficiency of women is on the whole inferior to that of men" and that this pattern was evident across the animal kingdom: male animals tended to be faster than their female counterparts. Yet these categories were not absolute. There was a great deal of "overlap" between adolescent girls and boys in terms of height, weight and speed with those girls at the peak physical fitness for their gender better than those boys who in the "lower third of the boys' sample". Further, environment played a role, and physical sports were historically male-focused and it was therefore "not fair" to apply the same standards to women. Simultaneously, in terms of cognitive tasks, young girls tended towards the domestic and maternal with young boys more adept at physical co-ordination and problem-solving. Again, these were not absolute categories but related to "statistical averages" and Jokl's analysis made allowances for women and girls to take on historically male activities—provided that they demonstrated that they were "fit" for them. The notion of "fitness" however was key, with his view that the "the scope of human fitness is limited" so no individual was ideally "fit" for all tasks. Yet fitness could be adjusted through the process of training where, with constant and sufficient exposure to a particular stimulus, the "organism"

[69] Daries, "The History of Physical Education at Stellenbosch University", Chapter 2.

would improve in its capability or "fitness".[70] This was as true for physical exercise as it was for the demands of flight—yet those demands were constantly changing and Jokl's research into aviation medicine suggested to him that, while all measures had to be taken in order to understand physiological processes and hence protect the body of the pilot, this body was ultimately to be subordinate to the machine. While he was willing to concede female capability, the body of the pilot was also a gendered one contextualized by the historical role of the aeroplane as a tool of power.

The development of the aeroplane and its evolving relationship with the male body were embodied in the practice of aviation medicine. With the early aeroplanes used during the First World War, flying was ultimately a militaristic and a masculine enterprise, permitting an elite to demonstrate their prowess and achieve individual glory that belied the industrialization and alienation of modern warfare. As aircraft became more sophisticated however, the male body became an addendum to the machine. Tests were designed to determine their suitability and procedures implemented to increase their fitness to prepare airmen for the increasing physiological demands of the speed and altitude that now characterized aircraft and their use in warfare. This increasing trend of subordinating the human being to the modern machine was briefly checked as the Second World War drew to a close. Towards the end of 1944, an increasingly desperate Japan began using kamikaze pilots—young men who, in the very antithesis of the aims of aviation medicine, used their aircraft as an expression of their will, turning both their bodies and the machine itself into a destructive—and self-destructive—weapon. The Germans would consider a similar strategy with their unfulfilled vision of the *Volkssturm*—teenage boys flying in "disposable" aircraft against Allied pilots, who could then use their aircraft to ram Allied planes when they ran out of ammunition. While the *Volkssturm* plan was not implemented, the destructive capacity of the kamikaze pilots was limited, with superior Allied technology used to intercept and for the most part, minimize their impact.[71] Technology ultimately held sway over the male body and will.

[70] Refer to Ernst Jokl "Medical Research in Physical Education in South Africa", 1947, reprinted in Jokl, *South African Reminiscences*.

[71] Shachtman, *Laboratory Warriors*, pp287–288, 289.

CHAPTER 8

The Measured Mind: Aptitude Testing in the Union Defence Force

In February 1940, Madame Ada Douvern was invited to give a lecture to the Johannesburg and District Sub-Centre of the Institute of Transport, the first woman to do so. Her lecture was on "Psychology in Transport Undertakings", and she was particularly enamoured with the way in which an understanding of the psychology of the client was used to boost American business. This could then be applied to the South African context and to public transport in particular, where knowledge of the different psychological types would result in better service and "create admiration and satisfaction in the customer's mind". For Madame Douvern, there were four kinds of people: "sanguine, choleric, melancholy, and phlegmatic", each of whom had specific attributes. Their psychological categorization could be manifest in their physical appearance as well. Those classified as "choleric" were not just "seeking improvement and advancement" and "radiated strength and power of leadership", they also tended to have "long eye-lashes, prominent nose or chin or perhaps forehead". Assertive and demanding, they were likely the most difficult to please.[1] Speaking just a few months into the Second World War, Madame Douvern's psychological classification drew heavily

[1] "Psychology in Transport Undertakings", *South African Railways and Harbours Magazine*, Vol 34, April 1940, pp493–494.

S. Chetty, *Science and Technology in South Africa, 1939-1946*, Palgrave Studies in the History of Science and Technology,
https://doi.org/10.1007/978-3-032-05084-7_8

upon theories of medicine promoted by the ancient Greeks more than two thousand years previously. Accredited to Hippocrates, the father of modern medicine, illness was attributed to an imbalance in the four bodily humours: blood, phlegm, black bile and yellow bile. These, too, had an impact on personality and race; northern Europeans were phlegmatic, Africans bilious, with the Greeks unsurprisingly embodying the harmony of the humours. A ubiquitous Western medical doctrine for millennia, the humours had fallen out of favour by the nineteenth century with the increasingly scientific basis for modern medicine and psychology—even if Madame Douvern remained unconvinced.[2]

It was, however, this new psychology that would underpin both South Africa's role in the War and its path to modernization.

The experiences of the First World War had proved a steep learning curve for the British. Charles McMoran Wilson (better known as Lord Moran) wrote the first account of the psychology of "shell shock" during the First World War. In a book replete with poignant case studies, Wilson described courage as a "moral quality" and as "will power". It was an attribute based on individual resilience. However, the conditions of the trenches, the fear, the regular confrontation with violent death, the effects of shells and gas—all hallmarks of the first industrialized modern war of the twentieth century—depleted the individual's stock of will power. It was a war of attrition with the constant "expenditure" of the individual's will until nothing remained. Yet individual attributes were also subordinate to military discipline and soldiers exposed to and traumatized by gas or shell fire—even with relatively minor physical injuries—were sent back to the trenches or executed for desertion with their actions perceived to be due to a lack of courage/will power/moral fibre. The discounting of psychological trauma compelled Wilson to pose the pained question: "Were they really unable or only unwilling?".[3]

Yet the notion of individual agency and culpability in relation to war trauma remained dominant during the interwar period. In 1922, the War Office Committee of Enquiry into Causation and Prevention of "Shell-Shock" considered the changing definition and increasingly widespread use of the term a misnomer. Initially suggesting the concussive and

[2] Roy Porter, *The Greatest Benefit to Mankind: A Medical History of Humanity from Antiquity to the Present* (London, Fontana Press, 1999), pp56–57.

[3] Lord Moran, *The Anatomy of Courage* (London, Constable and Robinson Limited, 1945), pp67, 69, 187–188.

subsequent psychological effect of being exposed to exploding shells, the changing demographic of the Army brought about by increasing enlistment as well as the implementation of conscription, led to "astoundingly numerous" cases of soldiers exhibiting similar symptoms that were unrelated to proximity to exploding shells. "Shell-shock" then came to be applied to those unable to effectively make the transition from civilian to military life and would subsequently display "neurasthenic and hysterical symptoms". It also referred to soldiers experiencing the horrors of trench warfare on the Western Front who went on to develop "war neuroses". The suggestion that increasingly widespread recruitment and conscription correlated with larger numbers of soldiers experiencing "shell-shock" suggested an individual failing, perhaps incurred by the level of "morale" of the soldier. Training was perceived to be the solution to reduce the possibility of "war neuroses" with an emphasis on "[promoting] morale, esprit de corps and a high standard of discipline". The soldier should learn to prioritize the well-being of his unit over "his own physical safety". Further, regimental officers and medical officers should co-operate in order to determine those most likely to have the "abnormalities from which mental or nervous instability may be inferred". Officers should also be trained in "the study of character". The relationship between "war neuroses", and "character" was further evident on the frontlines where "slight cases" were to be encouraged to return to combat and "No soldier should be allowed to think that loss of nervous or mental control provides an **honourable**[4] avenue of escape from the battlefield".[5] By the end of the War then, while there were attempts to treat and to prevent "shell-shock" or "war neuroses", these were also contextualized by the official view that these psychological disorders were a moral failing associated specifically with the lack of fortitude on the part of the individual. It could be applied to those who were perceived to be not fully committed to comrade and country.

The apparent lack of moral fibre manifest as the inability to engage in combat and was, as Joanna Bourke shows, linked to technology. On the one hand, the use of technology—the weapons of war—allowed for

[4] Emphasis added.

[5] "Committees (Code 18(C)): War Office Committee of Enquiry into Causation and Prevention of 'Shell-Shock'; minutes and correspondence report", 1922, WO 32/4748, The National Archives, Kew (hereafter TNA).

distance from the actual experience of killing. As evident at Fort Leavenworth during the Second World War, a counsellor observed that soldiers were so "preoccupied [with] physics and ballistics" that they neglected to consider the actual role of their weapons in the taking of human life. Further, the more powerful the weapon, the greater the "thrill of destruction". A veteran writing in the 1980s noted the way in which the modern machine gun was akin to the medieval sword—and an extension of the arm of the combatant. This was akin to the relationship between aviators and their aeroplanes. Conversely, during the First World War and particularly evident in the Second World War were anxieties about "passive combat personnel", those soldiers who had been trained, had not succumbed to "shell shock" but who nevertheless did not fire their weapons. Therefore, even when men were equipped with state-of-the-earth weaponry, they were nevertheless less inclined to use them in combat even when coming under fire. They were deemed to have a dearth of "the offensive spirit". This, too, was attributed to technology with the very weaponry associated with modern warfare, alienating the troops who wielded them. The distance of soldiers from each other on the battlefield, the distance from the enemy led to a sense of "isolation and confusion", making "the [industrialized] battlefield one of the loneliest places on earth", and thereby inhibiting the aggressive impulse.[6]

For South Africans, cases of "shell shock" were diagnosed from the Second South African War. Here, in accordance with the original definition of the term, the afflicted were those in close proximity to artillery fire during the Siege of Ladysmith where British troops in the town of Ladysmith were besieged by Boer forces for more than three months. During the First World War, despite the Allied use of "mental hygiene teams" to treat cases of psychological trauma incurred by war, the sheer scale of the War and its deleterious effects were unprecedented, and psychological services were not up to the task. For South Africans, there was very little capacity in the country to deal with psychological trauma and Union Defence Force (UDF) soldiers were treated in England and only sent home in 1920.[7]

[6] Joanna Bourke, *An Intimate History of Killing: Face-to-Face Killing in Twentieth-Century Warfare* (London, Granta Books, 1999), ppxvii, 2, 61–65.

[7] Warren Burgess, "Foreword", *Scientia Militaria,* Vol 50, No 2, v–xv, 2022, px.

By the outbreak of the Second World War, the use of psychology was integral to the mental fitness of the soldier in two ways. The first was related to the treatment of "war neuroses". In the interwar period, hospitals were set up specifically for the treatment of soldiers and civilians suffering from the debilitating trauma of war. The converse of treating those who had succumbed to the stresses of war was to determine the suitability and aptitude of military personnel so as to ameliorate the effects of "war neuroses" and to place military personnel in the positions for which they were psychologically most suited, and which would ultimately make more effective the fighting force. Linked to this were the demands of modern warfare where "even the infantry in these days cannot employ the dullest section of the normal population, so rapid has become the pace of the modern army, so highly is it now mechanized". Suitability could then be determined through an instrument of measurable scientific rationality: the aptitude or intelligence test. Testing could be used to determine those wholly mentally ill-equipped for modern warfare as well as those whose superior intellectual abilities could lead to advancement through the ranks or place them in increasingly skilled and demanding positions.[8]

With the naming of "shell shock" in 1915, the director of the National Committee for Mental Hygiene in the United States, Thomas Salmon recommended the "[rigid exclusion of the] insane, feebleminded, psychopathic and neuropathic" from war service. Upon the United States's entry into the First World War then, personality tests were administered to recruits to determine their suitability for the Western Front and to remove those considered to have "weak constitutions" and hence susceptible to neurological disorders. By mid-1918, there were a number of psychology and psychiatry military divisions as well as a school for the training of military psychologists. Recruits would undergo a "psychological examination" to "[eliminate the] mentally unfit". More than 1.7 million American recruits were given various types of personality tests, with two per cent of them failing to make the grade. These tests were also applied to the workplace to determine intelligence and aptitude. Early tests were rudimentary, posing a series of questions to which subjects had to respond in the affirmative or the negative. Later tests would build on these and by the 1930s, testing focused on complex personality traits such as "neurotic tendency,

[8] Charles S. Myers, "Applied Psychology in War", *Manchester Guardian* reprinted in *South African Railways and Harbours Magazine*, Vol 35, July 1941, p803.

self sufficiency [and] introversion".[9] By this period, the focus was more on prevention than on cure with less attention paid to the treatment of "shell shock". The emphasis was instead on the use of testing to exclude those most likely to develop war neuroses in combat. The dependence on preliminary psychological screening meant that recruitment officers were held liable if they permitted the recruitment of men who were deemed "mentally unstable".[10]

Beginning in April 1940, psychiatrists attached to the British Army began administering tests to determine the "innate intelligence" of recruits. The tests were to weed out those who were considered to have a "mental defect or low intelligence" and determine those recruits of higher intelligence that could be trained for positions of command. Just as important as determining those of below average intelligence was the need to prevent the development of "neurotic symptoms" on the part of men who had been placed in "less skilled occupations" than their intelligence suggested. From the beginning of the recruitment process then, "scientific selection" had to be applied to deploy men in positions for which they were most suited and hence make the army more effective. Recruits were thus given the Penrose-Raven Progressive Matrices test.[11]

Lionel Sharples Penrose was a geneticist whose work stood in opposition to eugenics thinking in post-First World War Britain. Eugenicists as exemplified by the Eugenics Education Society considered socially undesirable traits such as alcoholism and deviant behaviour to be a product of genetics—as was "feeble mindedness". Amidst post-war concerns of "national degeneration", eugenicists contemplated the forcible sterilization of people with "mental disabilities" so as to prevent a decline in population intelligence. Penrose's counter to the eugenicists was based on scientific thinking arising from his training in mathematics and medicine, and his conclusion was that mental disability was a result of a complex interplay between both environmental and genetic factors—and not solely

[9] Lila Thulin, "The First Personality Test was Developed during World War I", 23 September 2019, *The Smithsonian Magazine*, https://www.smithsonianmag.com/, accessed 6 October 2024.

[10] Annessa C. Stagner, "Healing the Soldier, Restoring the Nation: Representations of Shell Shock in the USA during and after the First World War", *Journal of Contemporary History*, Vol 49, No 2, 255–274, 2014, p269.

[11] Robert H. Ahrenfeldt, *Psychiatry in the British Army in the Second World War* (London and New York, Routledge, 1958), pp33, 35.

based on heredity. His realization came as a result of his research conducted in Colchester.[12]

Like Penrose, John Carlyle Raven was based at the Royal Eastern Counties Institution in Colchester, serving as a psychologist. The "Colchester Survey" of which both men were part was an investigation into the causes of "mental defect" and intelligence tests were administered to children diagnosed with "mental defect" as well as to adult family members. The test used was the Stanford-Binet intelligence test developed some two decades earlier. It was, however, as a realization of the shortcomings of the Stanford-Binet test that Raven found both tedious to administer and unreliable, that he developed the Progressive Matrices test. It was a test that could be applied to both adults and children, did not depend on literacy and could conveniently be conducted at home, school and work.[13]

With the outbreak of War in 1939, Raven—a conscientious objector—began conducting research into the validity of his test in determining recruit suitability. The test was given to all new recruits and a twenty-minute version of it was adopted by the War Office Officer Selection Boards to determine aptitude for command and leadership. As a test that did not depend on language, it was soon adopted by the other Allied countries—and would later make the transition from the military to education.[14] The value of the test lay in its adaptability making it suitable for anyone regardless of "age, linguistic ability, or physical defects". Those tested fell into five categories ranging from "Grade I or intellectually superior" to "Grade V or intellectually defective". The former were those with the requisite intellectual ability to pursue a tertiary education and similarly skilled occupations. Those classified as "Grade IV or definitely below average", on the other hand, were largely suited for unskilled work.[15]

[12] Maria Kiladi, "A Historical Perspective on Lionel Penrose: Scientist, Geneticist and Dedicated Opponent of Eugenics", *Annals of Human* Genetics, 1–8, 2024, pp1–3.

[13] John and Jean Raven, "John Carlyle Raven, 1902–1970, and His Legacy", http://eyeonsociety.co.uk/resources/J-C-Raven-Biography.pdf, accessed 5 October 2024, pp11–12.

[14] Raven, "John Carlyle Raven", p12.

[15] John C. Raven, "Standardization of Progressive Matrices, 1938", *British Journal of Medical Psychology*, 19, 137–150, 1941, pp137, 145.

The intelligence testing approach to recruitment and the subsequent assignment of soldiers did not necessarily meet with unanimous approval with the more recalcitrant "senior officers" believing that men could manipulate the system so as to avoid military service. In addition, for these officers, the ideal fighting soldier was in fact the "dull [man]", unquestioningly following orders. For the old guard, the military recruits required for "modern" warfare was no different than those of times past. Yet, for the most part, the testing initiative had the support of those who had been disillusioned with the "arbitrary methods" used to "[allocate] personnel". The testing system—with its suggestion of quantitative scientific rationality—indicated the ideal placement of the right man for the right job.[16]

While certain elements of the British Army made the claim that there was little required that was intellectually challenging for the soldier regardless of the notion of "modern" warfare, the need for selection in the Army was arguably greater with the bulk "of [the] intelligent and able men" taken up by the Navy and Air Force, thereby leading to an "[accumulation] of men whose mental background was questionable" in the Army.[17] Yet the converse was also true. The demand for skilled men of above average intelligence was just as essential in the Navy and the Air Force with their use of complex equipment and machinery. In the United States, for instance, the USS *New Jersey* was an Iowa-class battleship conceived in 1938 and launched on 7 December 1942. Bristling with anti-aircraft guns, the ship was also equipped with air and surface radar detection and entered the record books as the "fastest battleship in history". The ship would be used to great effect in the Pacific theatre of war—and for the later American conflicts over the course of the twentieth century.[18] In 1943, this state-of-the-art vessel had a crew complement of 2,600, most of whom had little sea experience. It was in this context that the National Defense Research Council provided the expertise of its Applied Psychology division to test and train the novice crew. The ship became a model for the idealized collaboration between psychologists and the military. For the latter, the psychological expertise had created

[16] Ahrenfeldt, *Psychiatry in the British Army*, p36.

[17] Ahrenfeldt, *Psychiatry in the British Army*, p32.

[18] "Full History—USS *New Jersey*: The World's Greatest Battleship", https://www.battleshipnewjersey.org/, accessed 5 October 2024.

an effective crew and for the psychologists, their field based on testing and "measurement" had now acquired the characteristics of a "'hard' science".[19]

In terms of psychological testing and training as it applied to the UDF, the South Africans were not as favourably placed as their British and American allies. The dearth of experts meant that testing was only applied to the newest branch of the UDF, that which most exemplified modern warfare.

South African Air Force (SAAF) pioneer, Pierre van Ryneveld had enlisted in the Royal Flying Corps in July 1915 as a pilot. With the end of the First World War, he was recalled to South Africa by Jan Smuts who named him the Director of Air Services tasked with forming South Africa's own military air division. Van Ryneveld, then, became the "father" of a separate branch of the UDF, the SAAF.[20] In addition to applying his First World War experience to the formation of the SAAF, Van Ryneveld had also been exposed to the methods used to select French pilots during the First World War and, as eventual Chief of the UDF in 1939, he initiated the formation of a similar unit for the testing and selection of South African pilots. This would become the Aptitude Test Section (ATS) under Simon Biesheuvel.[21]

Pilots' experiences of war were mediated by their aeroplanes. Writing of American pilots during the Second World War, naval psychologist Chester Bennett claimed that their form of combat was part of a "context of mechanical skill and strategy", a context which pitted aeroplane against its other or against distant targets on the ground.[22] Yet, as Wilson recognized from as early as the preceding war, there were specific kinds of psychological stresses placed on pilots that could be distinguished from the infantry in the trenches. In the anonymity of modern warfare, the pilot retained individuality, yet it was one tempered by his isolation in his machine—and the same applied to bomber crews. As squadrons took part in bombing campaigns, they communicated with each other via radio

[19] Tom Shachtman, *Laboratory Warriors: How Allied Science and Technology Tipped the Balance in World War II* (New York, Perennial, 2003), pp233–234.

[20] Kirk Kinnear, "Father of the SAAF", 26 December 2010, https://web.archive.org/web/20111004201645/http://www.saafmuseum.org.za/features/2010/father-of-the-saaf/, accessed 23 September 2024.

[21] Burgess, "Foreword", pvii.

[22] James Bradley, *Flyboys* (New York, Back Bay Books, 2004), p179.

to alleviate the isolation as they flew the long distances to their targets. Pilots and their machines were also particularly susceptible to the vagaries of weather: fog that reduced visibility or the absence of cloud cover that made them easy targets for anti-aircraft fire. Most evident was their utter dependence on the machinery and equipment that symbolized modern warfare. It is ironic that, while the ability to operate sophisticated and complex machinery such as aeroplanes meant that pilots were drawn from those who tended to score highly on aptitude tests, the possibility of mechanical failure—often beyond the pilot's control—created a great sense of anxiety.[23] It is unsurprising then that, just as Ernst Jokl addressed the physiological requirements for pilots in terms of dealing with altitude, g-forces and oxygen deprivation,[24] Simon Biesheuvel focused on aptitude testing to determine those most suited to the intellectual and psychological requirements of flying.

> It is obvious that an Airman should possess courage and determination, resistance to fatigue, be capable of eliminating disturbing thoughts from his mind, and possess qualities of dash, initiative and leadership.[25]

So wrote Captain Macfarlane, the Acting Office in Charge of the Aptitude Test Section of the No. 75 Air School. This was borne out in a recruiting supplement issued by the UDF in June 1940 that emphasized the youth of SAAF candidates, young men able to meet the challenges of training and of air combat who exhibited "clear and calm judgement" and were able to "[make intelligent] decisions".[26] It was at No. 75 Air School where recruits were tested for the required aptitude and temperament and subsequently deployed as a "Pilot, Observer, Wireless Operator or Air Gunner", assignations that Macfarlane believed to be done with "unfailing accuracy".[27] No. 75 Air School in Lyttelton just outside Pretoria was labelled "the kindergarten of the Air Force". The school had been established by Colonel C.E.S. Bull, with the Officer Commanding

[23] Moran, *The Anatomy of Courage*, pp104–105.

[24] Please refer to the previous chapter focusing on aviation medicine.

[25] Captain E. Macfarlane, "Psychology in the Air Force", *The Nongqai*, July 1943, p649, DOD.

[26] "The Union at War: How to Join the Forces", *The Nongqai*, June 1940, p654, DOD.

[27] Macfarlane, "Psychology in the Air Force", p649.

Major P.A. Moore and the Chief Instructor Major N.J. MacRobert. Also attached to the school was Lt-Colonel Simon Biesheuvel. An exception to the conventional flight school in that it did not carry out practical training, No. 75 Air School was instead the site where ability was determined. Potential recruits were given a medical examination and upon being pronounced "physically fit", were sent to No. 75 Air School for a barrage of psychological and psychometric testing. Upon completion of the tests, candidates were categorized according to the skills for which they appeared most suited, from pilot to air gunner and went on to begin their training. So successful was the determination of aptitude that Biesheuvel would be invited to London by the British Air Ministry, there to explain the testing process developed in South Africa—at his behest.[28]

The man who would be described in his obituary as "the doyen of [South African] psychology" was born in the Netherlands in 1908. His father's career meant that the family moved frequently, and they were in Belgium during the outbreak of the First World War. The German invasion forced the family back to the Netherlands. With the end of the War, there was a brief sojourn in Brazil followed by a return to Belgium before the Biesheuvel family emigrated to South Africa. By the time the fourteen-year-old Simon moved to Cape Town, he had experience of a wide range of cultures and was fluent in several languages—which would place him in good stead as he developed an interest in human personality. He majored in Psychology and English at the University of Cape Town; this was followed by an MA and he was awarded his doctorate in Psychology from the University of Edinburgh in 1933.[29] Returning to South Africa, he would take a special interest in intelligence testing.

From as early as 1916, intelligence testing was conducted on black and white South Africans yet there remained uncertainty as to whether measured differences were based on the stark distinctions in terms of education as well as other environmental factors. At the same time, as Dubow shows, Charles Loram conducted intelligence testing on white, Indian and African children, ranking them in a racial order of aptitude that would buttress segregationist thinking. However, on the eve of the Second World War, a more liberal perspective was evident with researchers

[28] H.P.H. Behrens, "Seventy-Five Air School: the Kindergarten of the Air Force in South Africa", *The Nongqai*, July 1943, pp641–642.

[29] J.M. Verster, "Simon Biesheuvel, 1908–1991", *South African Journal of Psychology*, Vol 21, Issue 4, 267–270, 1991, p267.

acknowledging nurture, while not necessarily discounting nature. Wits psychologist, I.D MacCrone criticized the use of intelligence testing to support the notion of inherent racial difference and inequality, considering those who did so to be using science to support ideology or "prejudice". His ire was aimed in particular at the work of figures such as M. Laurence Fick who wrote of a "marked inferiority" of the "Native" when compared to their white counterparts—an intellectual inferiority not solely a product of environment but on inherent difference.[30] This was by no means a purely South African phenomenon. As William Percy shows, the interwar period in the United States was marked by the exclusion of black men from the Army Air Corps, both for reasons of segregation but also the findings in studies conducted by the Army War College that black men "lacked the mental and physical capabilities to make an effective soldier". It was only in 1940 that legislation permitted the formation of black combat units, although these remained segregated. It was in this context that black Army Air Corps units were formed in Tuskegee, Alabama, with their wartime success countering prevailing views of black inability.[31]

Simon Biesheuvel's *African Intelligence* was written in part as a response to Fick's *The Educability of the South African Native* published in 1939. For Biesheuvel, intelligence as measured by testing, was not simply a measurement of "innate ability" but a result too of environmental factors. Fick's conclusion that there were innate differences in intelligence between "native" and "European" children had therefore neglected the influence of the environment. For Africans living in urban areas, attention was drawn to the small homes, "squalid" conditions and the instability wrought by migrant labour on the family. There were also marked disparities in the provision of education for black and white children. Further, the "[inferior] standard of living" of the black population when compared to their white counterparts resulted in a poor diet observed in the former with a concomitant effect on the development of intelligence. By addressing the influence of environment—rather than perceived "innate ability"—Biesheuvel criticized Fick's findings that

[30] Saul Dubow, "Mental Testing and the Understanding of Race in Twentieth-Century South Africa", *Science, Medicine and Cultural Imperialism*, Teresa Meade and Mark Walker, eds (Basingstoke, Macmillan, 1991), pp152–153, 155–156.

[31] William Percy, "Jim Crow, Uncle Sam, and the Formation of the Tuskegee Flying Units", *Social Education*, Vol 63, No 1, 14–21, 1999, pp14–15, 18.

Africans were only suited to a "rudimentary" education and that African "inferiority" was immutable. Echoing MacCrone, both conclusions were considered by Biesheuvel to be a result of "political bias or emotional needs", a case of placing ideology before scientific reason.[32]

And the scientific method was the cornerstone of his work during the Second World War. Biesheuvel was 32 when he enlisted in the UDF and was from early 1941 attached to the Aptitude Test Section (ATS) of the SAAF.[33] Given the lack of South African resources at the outbreak of the War, it was essential to apply training to those most suited for it—which was the rationale behind the formation of the ATS in 1940, comprised largely of psychologists drawn from the University of the Witwatersrand (Wits). The ATS was tasked with determining most efficiently those potential airmen who demonstrated the intelligence and personality that suited them for the stress of combat flight, or what was termed a "Service Temperament". Once identified, they underwent additional tests to determine which area of combat flight they should be assigned: "fighter duties; bomber duties; reconnaissance flying; flight instruction". In time, the role of the ATS was expanded to include the assessment of other Air Force personnel such as those involved in communications, engineers, technicians, maintenance and ground crew. The ATS would also look at those who had "failed" or who had become "ineffective"; conduct research into the stress associated with flight as well as "accident proneness" so as limit "air accidents"; and look for means of making more efficient the training process. Biesheuvel was eventually part of a staff that comprised eighty-five personnel ranging from psychologists to those tasked with designing and constructing the unique apparatuses in the workshops as well as administrative staff.[34]

Biesheuvel—who in civilian life lectured Psychology at Wits—had to adapt his understanding of aptitude to the requirements of flight. To do so, he, along with other members of the ATS, took flying lessons and also had discussions with flying instructors whose experience had given them an understanding of the reasons for "wash-outs and other flying

[32] S. Biesheuvel, *African Intelligence* (Johannesburg, South African Institute of Race Relations, 1943), pp18, 20, 21–22, Chapter IX, 128–129, 154.

[33] "Biesheuvel, Simon", Service Card No W.R.12/7542, 203333V, DOD.

[34] Simon Biesheuvel, "Air Personnel Research in South Africa", Box I4: Notes on Various Topics in Psychology, Professor Simon Biesheuvel Collection, 1933–1991, WaB-02-Bie-S, University of the Witwatersrand Central Records Office (hereafter WitsCRO).

problems". It was also necessary to understand the particular stresses of flying under combat conditions—which was not the same as that of taking instruction on the home front. The ATS also travelled to the Middle Eastern theatre, observing the flight crews actively involved in bombing raids so as to determine both the "stresses" they experienced as well as the "qualities" required in the ideal pilot. In what Macfarlane described as an application of the scientific method of experimentation and observation, Biesheuval subsequently isolated the relevant requirements for flight, and these became the basis of a series of tests administered to recruits. The emphasis was on determining those most suited to be pilots with the understanding that a test to weed out the unsuitable was less time-consuming and more cost-effective than flight training.[35]

Broadly, tests consisted of those measuring "mental ability and skill" as well as "temperamental and personality qualities". Included too, was a consideration of the individual's own "interests".[36] Skill could be subdivided into "higher mental skills" as evident in the measurement of intelligence, rational thinking, the ability to remember, verbal and numerical capability. "Perceptual skills" were those related to sensory input and also to the ability to situate oneself spatially—a necessary skill for flight. "Motor skills" implied co-ordination and manual dexterity. Temperament and personality were individual attributes that were a challenge to determine, yet no less important—introverts, for instance, were not suited for group work while extroverts potentially lacked the ability to focus on a single task for long periods of time without stimulation. The determination of temperament and personality was also complicated by the subject's own possible lack of self-reflexivity. The testing process here addressed three aspects of temperament and personality: "constitutional attributes" focusing on behaviour, the need for stimulation and ability to focus; and "acquired attributes" which addressed one's capacity to adjust and function in various social situations. Finally, there was "character make-up" which considered the individual within the context of social values and expectations; the extent to which the subject would be able to adhere to "ethical" norms and be "conscientious…in pursuit of duty". Like

[35] Macfarlane, "Psychology in the Air Force", p649 and Captain J.O. Rowe, "Psychological Tests for Air Crews", *The Nongqai*, July 1942, p804, DOD.

[36] Simon Biesheuvel, "Scientific Personnel Selection", talk delivered on 8 February 1952, Box I4: Notes on Various Topics in Psychology, Professor Simon Biesheuvel Collection, 1933–1991, WaB-02-Bie-S, WitsCRO.

temperament and personality, the measurement of "interests" was also a challenge. This was idiosyncratic, based on the "interest" and hence the motivation of the individual, where someone with a keen interest in their task would be more likely to persevere even if there were setbacks, than someone with a lack of interest and motivation.[37]

Depending on type of assessment, tests could be administered either to individuals or to groups with the former more suited to the testing of personality and the latter used for "mechanical" aptitude. Key to all tests administered however was standardization where instructions and other conditions remained constant for all candidates.[38] The tests administered ranged from the simple to the complex and in 1943, staff at the UDF publication, *The Nongqai* were taken on a tour of a testing laboratory with "room after room...containing unusual and interesting tests". All equipment for the tests had been designed and built in the laboratory workshop and were the product of collaboration between the "Psychologist and the Technicians", in addition to consultation with the experts at Wits. This alliance between science and technology produced machinery that was unique, and devices were connected to recording equipment so that graphic and statistical measurements were made determining a recruit's ability under both normal and "stress conditions". *The Nongqai* team's interest was piqued by a number of devices designed to test co-ordination. The first dealt specifically with the co-ordination of hands and feet and comprised a glass screen upon which a moving spot of light was projected. The candidate was expected to use a "rudder bar" (such as that found on a plane) in order to return the light to its "central position". Another device which operated similarly focused on eye-foot co-ordination. Another test related to determining "reaction time and...ability to learn". Here four switches were connected to four lights on a board and the recruit had to determine both the light and its controlling switch and then "operate the lights in a particular sequence" with his speed and accuracy subsequently measured. A test known as the "Perseverameter" measured the "mental inertia" evident when a prospective pilot was given a new task, the skills of which had partially been learned in a previous task. It was used to determine the ability of pilots to learn

[37] Simon Biesheuvel, "The Scope and Purpose of Aptitude Tests", Box I4: Notes on Various Topics in Psychology, Professor Simon Biesheuvel Collection, 1933–1991, WaB-02-Bie-S, WitsCRO.

[38] Biesheuvel, "The Scope and Purpose of Aptitude Tests".

to fly new aircraft. Candidates usually had the requirements of the tests explained to them beforehand and were given time to first acquaint themselves with the devices. During the test, each attempt took approximately one minute. And, just as prospective pilots were being tested, so too were the testing devices and the tests were constantly undergoing a process of refinement. Finally, there was the "Link Trainer", an early type of flight simulator that tested all the skills required for piloting an aircraft, while later serving as a training device to acquaint pilots with flight conditions.[39] Biesheuvel was remembered as being an archetypal "boffin", bent on subjecting candidates to a battery of tests that included the use of a "decompression chamber" termed "Biesheuvel's Baby". Prospective test subjects/pilots apparently "[scattered] like frightened deer" at the sight of him.[40]

The barrage of tests was designed to elicit and measure "reactions" which could be classified. These reactions included compliance in terms of taking instruction, questions to test knowledge, the ability to distinguish sensory input such as different levels of sound or different visual stimuli, the testing of reasoning and "problem solving", the rate at which tasks were accomplished and the quality of work. These reactions were measured differently with some tests being multiple choice while those requiring an essay-type answer were assessed according to an established standard. "Interests", for example, were determined through the use of a "questionnaire" that focused on aspects such as "activities, hobbies and sports" as well as the candidate's biographical history, all of which were correlated with results from the various skills tests such as the IQ (Intelligence Quotience) test. It also had a gendered component with "a masculinity-femininity scale".[41]

"Professional testers" were assigned to observe the candidates as they carried out the tests. Observations made related, among other factors, to precision, perseverance—even when confronted with failure—and level of focus. These were then recorded on standard forms using a rating system. The testing process followed a scientific method: candidates were unaware

[39] Macfarlane, "Psychology in the Air Force", pp650–651. The Link Trainer is discussed in greater detail in the previous chapter on aviation medicine.

[40] Burgess, "Foreword", pvii.

[41] Biesheuvel, "The Scope and Purpose of Aptitude Tests".

of their observation and testers had been vetted and their "biases" determined and hence taken into account in their assessment. At the end of the process, temperament was rated in terms of "stimulability of the personality", the extent to which emotion affected performance and the "drive" of the candidate. Using a "five-point scale", the tester would then recommend the most suitable position for the candidate.[42]

With the experience of members of staff who went to great lengths in terms of research and practical training to understand the psychological traits necessary for flight, Biesheuvel's team were able to further refine the types of "temperament" required for the different forms of combat flying. They concluded that temperament was related to the propensity for having an accident in the air; there was a marked distinction in temperament between those flying fighters and those who formed part of the crew of a bomber; that, even though a pilot may have performed well in "operation duties", this did not necessarily translate to being adept at combat flight. In addition to temperament, was mental aptitude and physical co-ordination—all of which were tested for, as discussed previously.[43]

The testing for the less measurable characteristics such as temperament and the reactions evoked was also context dependent. As Biesheuvel pointed out, attitude was key where an eager, young recruit was more likely to respond "normally" than a seasoned pilot, low in self-confidence with high anxiety levels. These differences were also evident in tests given before combat and the results obtained after combat—and the stress associated with it. Context then limited the ability of these tests to serve as predictors of behaviour, thereby affecting adversely their scientific validity. The same applied to "character tests" which were also criticized by Biesheuvel with the "Test for Values" inadequate in determining "integrity and strength of will". With the difficulty in measuring these attributes, he recommended instead the use of interviews in conjunction with the various other tests in order to determine "character". The complexity of the process meant that "character" determination should ideally only be taken when it came to leadership or "higher grade" positions.[44]

[42] Biesheuvel, "Air Personnel Research in South Africa".

[43] Biesheuvel, "Air Personnel Research in South Africa".

[44] Biesheuvel, "The Scope and Purpose of Aptitude Tests".

The ATS had also been set up in haste. Given the demands made on the unit, methods of testing had to be developed and implemented with little delay. As such, Biesheuvel noted that they were unable to follow an idealized "academic" model where, following testing and assignation, the pilot would undergo training. There would then be a "follow-up" after he had concluded his first "operational tour". Only then could the efficacy of the tests be determined. The entire process would have taken at least three years which was not feasible during a time of war. By Biesheuvel's own admission then, the testing process was ad-hoc, a "gamble" regarding the "efficacy" and the "validity" of the testing process, underpinned by the reasoning that even a flawed selection process was better than none at all.[45]

The potential flaws notwithstanding, Biesheuvel was confident about the relevance of aptitude testing. Drawing upon the example set by the United States Air Force, he considered the ability of aptitude testing to predict success, with those passing the tests having a 95% chance of succeeding in their chosen field. In the South African context, he estimated that the correct administering of aptitude testing "nearly halved wastage in flying training". Further, by considering personality as an important component, candidates could be allocated and trained for the positions for which they were most suitable, thereby "reducing the number of cases of people unable to withstand the strain of operational flying".[46]

The first group tested comprised seven hundred men who had already enlisted, completed their basic training and were scheduled to begin early flight training. They had therefore not undergone the testing process at recruitment as this had yet to be implemented, and many had already been weeded out. The aim then in testing this group was to gather "follow-up data" once they had completed their training. The second testing group followed what would become the "normal" route—although testing had yet to be made compulsory. The early results suggested the efficacy of the tests. Even with poor results, candidates continued to the training process where it was discovered that those who tested poorly were more likely to be unsuccessful. Convinced by the predictive nature of the testing process then, the Director General of the Air Force made testing compulsory in

[45] Biesheuvel, "Air Personnel Research in South Africa".

[46] Biesheuvel, "Scientific Personnel Selection".

May 1942. This however also meant that it proved impossible to comprehensively check the "validity" of the testing method as only those who performed well went on to flight training, thus the failures could not be assessed as to whether they would actually be successful in training and as airmen. Another check on validity was the application of the same test to the Rhodesian Air Training Group (RATG). The RATG comprised Rhodesian cadets as well as those from the Royal Air Force that included cadets from Britain, India as well as southern Europe. Despite the diverse backgrounds, the assessments proved as valid in predicting success and failure—although the rating system was different.[47]

As scientifically valid as the tests attempted to be, there still remained the notions of strength of character and of moral failing that had been linked to "shell shock" in the previous war. An article appearing in *The Nongqai* in July 1942 a month after the surrender of the Second South African Division at Tobruk, described the experiences of a candidate who was determined by testing to have been "emotional and lacking in persistence". He went on to continue his pilot training but "complained of increasing tension and nervousness in the air, which led to insomnia and loss of appetite". Investigation into his upbringing revealed that he was from an affluent background with a father who indulged his every whim. As a result, he found the rigours of military training "soul-destroying", claiming that it "had broken his spirit". He was subsequently declared a "voluntary washout".[48] It is not difficult to contextualize this vignette within concerns for the lack of moral fibre that had contributed to a defeat from which the UDF was still reeling. Telling however, is the belief that, had he been deselected based on his personality traits as revealed by psychological testing, it would have been of benefit to both the SAAF—and its resources—and the candidate himself who was inherently ill-suited for the position he desired.

Military service however, particularly during the context of war, inevitably took a toll on the serviceman.[49] For those in combat—or in proximity to combat—fear was a significant stressor. And, as Biesheuvel pointed out, "Fear is a delicate subject to discuss in a military context,

[47] Biesheuvel, "Air Personnel Research in South Africa".

[48] Rowe, "Psychological Tests for Air Crews", p804.

[49] It is important to note that Biesheuvel addressed servicemen and not women here—with an emphasis on those engaged in combat or experiencing the stress of frontline working conditions.

because of its intimate relationship with character and duty". Yet, unaddressed, its repercussions were potentially enormous due to its effect on "individual and group morale", making it a more dangerous "weapon" than that wielded by the enemy. The opposite of fear was of course "courage", and he drew upon the writings of Lord Moran and some of the seminal fictional depictions of war including Stephen Crane's *The Red Badge of Courage* to describe courage not as the "absence of fear but the ability to control it". A "breakdown" then could not necessarily be attributed to lack of moral fibre or a deficiency in "character". Instead, fear was described as "normal", a product of both instinct and environment. It evoked both a psychological and physiological response in the individual and, so long as it was "understood", the afflicted would be able to "withstand frequent attacks of it, without having his efficiency impaired".[50]

Yet how one experienced fear and the level of fear felt was based on individual attributes. These included "temperament" with those more "emotional" or "imaginative" expected to experience more acute fear. Those who lacked self-confidence and suffered from anxiety also tended to be more fearful and "[anticipated] failure". While fear was a biological and evolutionary mechanism, Biesheuvel also related it to environment where the characteristics associated with high levels of fear were derived from childhood experiences and the individual's interaction with parents and family. And just as the experience of fear differed across individuals so, too, did the ability to overcome it. Those used to feelings of fear may actually have been more likely to develop the skills required to function despite it. If one was isolated or had little "group support" or even experienced "fatigue", then one's ability to overcome fear was greatly lessened. For Biesheuval, it was a fallacy to believe that it was simply a matter of "will" as will was related to individual attributes of "temperament" and "character" constructed over a lifetime. Nevertheless, he was convinced that there were those who wilfully gave in to fear and it was those who "[deserved] moral censure for failure to fulfill their duties". There is simultaneously the acknowledgement that war heightened the experience of fear, far more than would likely be encountered in civilian

[50] Simon Biesheuval, "The Psychological Care of Combatants", Box I4: Notes on Various Topics in Psychology, Professor Simon Biesheuvel Collection, 1933–1991, WaB-02-Bie-S, WitsCRO.

life, while at the same time reducing the outlets and responses available to the individual confronted by the debilitating emotion.[51]

In combat—and in the military itself—individual differences were subordinated to the needs of the group, as was individual character. Group dynamics were essential, and an individual had the potential to be a negative influence on group "morale", thereby undermining military cohesion as a whole. Those individuals then had to be "eliminated by careful selection". If already part of a combat or other unit, and regardless of the environmental and childhood influences affecting temperament, Biesheuvel advocated for disciplinary measures for any individual who was remiss in "[carrying] out his duties...whatever the cause of this failure may be". An understanding of failure was not a justification for it as it was "not possible to build an orderly society...least of all maintain discipline in a combatant unit" should there be tolerance for the vagaries of fear. The sanctions recommended depended on the extent of the wilful nature of the offence and ranged from being "relieved of the duties with which he cannot cope" to being "punished" according to the strictures of military justice.[52] Biesheuvel's writing then demonstrated the same ambivalence regarding character that had been evident in early assessments of "shell shock", albeit his empathetic stance and attempt to contextualize those both afflicted, and rendered impotent, by fear. During a time of conflict, the interests of the group supplanted those of the individual.

Based on his experience in the ATS, it was the SAAF and the stresses of combat flying that illustrated best both the experience of fear and the overcoming of it. According to Biesheuvel, flight acted as a "high magnification microscope" that emphasized already existing "flaws in personality". This was due to the nature of fear evoked by flying—the usual anxiety related to combat was compounded by the very basic and instinctive "fear of falling". His work in the ATS then provided him with the scope to investigate the interaction between "confidence" as well as the various factors that impacted upon levels of fear and the resilience to overcome it. The relevance of the work was that it led only to the training of those determined to be suited to the particular stress associated with flight; it allowed the SAAF to put in place measures to address

[51] Biesheuval, "The Psychological Care of Combatants".

[52] Biesheuval, "The Psychological Care of Combatants".

"fatigue" and possible drop in morale; and it resulted finally in the development of methods to treat "breakdowns" which occurred "despite all precautions".[53]

With the end of the War, Biesheuvel was involved in assessing military personnel for post-war work. As an appendix to his report on fear and its treatment, Biesheuvel cited a number of case studies that were sent to him for consultation regarding occupational suitability in both the military and civilian spheres. "Lieut. G.F.", for instance, could either return to his pre-war job as a bus driver or, as he stated, train as a "pharmaceutical chemist". He was afflicted however with "black-outs" and Biesheuvel's assessment that, while he was intellectually suited to the position of a chemist, he was also very "predisposed to affective disorder…unstable and given to depressions". The conclusion then was that he would be suited to neither being a bus driver nor a chemist, but could instead be trained for administrative duties and find employment at the Pretoria Municipal Council where the "security" provided by the job would obviate his "tendency to a break-down". A subsequent case "S/Sgt. C.K." experienced high levels of anxiety in what could conceivably be considered a less stressful job as a "pay clerk" in the UDF. He was unable to cope, suffered a "break-down" and had to be reassigned. Biesheuvel's assessment of "C.K." addressed his childhood experiences which ultimately created a debilitating "fear of failure", and the recommendation was that he would not be suited for a supervisory role and that he be assigned tasks that were routine and mundane and on which he could focus constantly. In the case of "C.K." his intelligence was above average so there was little that he was incapable of doing however his psychological make-up limited the opportunities open to him. In contrast "Trp. W.O." who had already been treated at the UDF's psychiatric hospital, Tara, for alcoholism, was wholly unsuited to clerical work. His intelligence was fairly average and, while he had good language skills, his numeracy and mechanical skills were well below par. In addition to that which could be accurately measured, were the more nebulous factors of personality and temperament. "W.O." suffered from a lack of confidence and a "defeatist attitude" that could nevertheless be overcome with regular encouragement and praise. It was recommended that he be given a supervisory role in fields as diverse as "soil conservation"—where he

[53] Biesheuval, "The Psychological Care of Combatants".

could instruct farmers—to "native compounds". The feeling of responsibility, his language skills and, in the case of the latter role, "the lower status of the natives...[where] his ego will not feel challenged" would allow him to work independently, develop a feeling of confidence and thus remove the "temptation" of turning to alcohol. As such, he had the opportunity to "gradually build up his character"—even if Biesheuvel was sceptical about his success in doing so.[54]

The case of "Pte. A. v.B." was even more daunting. Described as being of "rather low mentality", "mediocre" aptitude and an "unimpressive" temperament, Biesheuvel rather hesitantly suggested opportunities in semi-skilled work or even unskilled work as a "porter or cleaner". Revealing in his assessment is his qualitative analysis of temperament as "unimpressive" on which he elaborated to mean "apathetic, self-conscious and nervous". Unlike intelligence testing then, which could be more clearly measured, personality and character were not so clearly defined yet adjectives such as "unimpressive" nevertheless suggested a moral-based assessment. This was evident too in the "building up" of "W.O.'s" "character" in the preceding case study. Biesheuvel's work thus related to the rendering of the individual suitable for assimilation into society and, while differences in character and temperament were acknowledged, they were nevertheless positioned against a particular societal standard—and found wanting.

"Tpr. D.E." presented another challenge—and one that had been evident in the First World War: hysteria. From antiquity hysteria was associated with women and—in a manner which Madame Douverne would have recognized—was attributed, as with other mental illnesses, to the humours being out of balance. Later, female hysteria would be believed to be due to a "wandering uterus" with marriage the "cure". By the end of the nineteenth century, Sigmund Freud, had a different concept of hysteria—although it remained as associated with the female mind and with "neuroses". For Freud, hysteria was based on the Oedipus complex related to "infantile sexuality" and attachment to the mother with a concomitant fear of the father.[55] This is reflected in Biesheuvel's analysis of "D.E." who was described as having been "[brutalized by] his

[54] Biesheuvel, "The Psychological Care of Combatants".

[55] Roy Porter, *The Greatest Benefit to Mankind: A Medical History of Humanity from Antiquity to the Present* (London, Fontana Press, 1999), pp81–82, 514–515.

father" and thus "wished the death of his father"—who would subsequently die some years later. While being described as being in a "normal" physical relationship with his wife, "D.E." also penned "phantasies" depicting an ambivalent view of women with one entitled "Die Brandstapel"[56] depicting the "ceremonial burning of women". His anxieties were also evident in his dreaming, the subject matter of which periodically included "assault, often by natives". Given to fainting spells, phantom pains and moments of "blindness" especially when he was serving in North Africa, Biesheuval diagnosed him with "hysteria", with psychotherapy recommended as a treatment.[57] The diagnosis of hysteria was one that originated in the "neuroses" of the First World War where hysteria—rather than being related largely to female maladies—became associated with male soldiers. While the British used terminology such as "shell shock", according to Marc Roudebush, the French employed the terminology associated with hysteria to describe similar male "neuroses". For the French, however, as with the Germans and the British, this was also depicted as a kind of moral failing, a lack of "resilience". Hysteria was moreover seen as a contaminant that could spread through the ranks, affecting morale, and therefore had to be contained. It was nineteenth-century French neurologist, Jean-Martin Charcot who took to applying hysteria more widely as a disorder that applied both to men and women. Charcot's student Joseph Babinski considered hysteria to be defined by its "false symptoms"—akin to the symptoms experienced by "D.E." where his "aggravated backache" was related to his anxiety and had no physiological cause. The significance of a "false symptom" was that its alleviation was then attributed to individual "will" that could be cured by "persuasion". Initially disregarded by psychologists, there was a resurgence in Babinksi's form of analysis during the First World War along with the ambivalence towards hysteria which was often perceived to be a form of "malingering". The treatment then for both "genuine" and "simulated" hysteria was the same.[58] While treatment and perceptions of hysteria had become more nuanced by the time of Biesheuval's

[56] Translated as "The Pyre".

[57] Biesheuval, "The Psychological Care of Combatants".

[58] Marc Roudebush, "A Battle of Nerves: Hysteria and its Treatments in France during World War I", Mark S. Micale and Paul Lerner (eds), *Traumatic Pasts: History, Psychiatry and Trauma in the Modern Age, 1870–1930* (Cambridge University Press, 2001), pp253–254, 261–263 and Biesheuval, "The Psychological Care of Combatants".

writing, the case of "D.E."—and of psychological assessment in general—nevertheless exhibits this same ambivalence between mental illness and the will of the individual, with testing the means of predicting and thereby controlling this uncertainty.

Yet there were cases where Biesheuval believed that testing could serve no use. "A/Cpl A.B." had been a nurse before the War and was referred to him with a case of mastitis (or pain in the breast) and anxiety. The patient, however proved to be manipulative, using both psychological and physiological symptoms to further her own interests. Biesheuvel conceded that there was little that could be done as her "maladjustment" would continue regardless of her field of occupation. Here, there seemed little doubt of "malingering" with his assessment of her character as displaying "psychotic features".[59]

In a talk delivered after the War, Biesheuvel reflected on the value of aptitude testing especially as it was used to set a benchmark for "human nature [which was] most resistant to the attempts by scientists to reduce it to law and order". Aptitude testing sought the general—a challenge when this had to be applied to the individual. Yet he remained convinced that testing—while still in its infancy—could be used to predict ability for particular types of work. Given the uncertainty of prediction however and the complex variables that related to both nurture and nature, aptitude testing worked with "probabilities" where prediction could be made to a degree of accuracy. As such, he addressed the criticism aimed at aptitude testing that had circulated in the form of largely apocryphal rumours: the candidate who had been rejected by the ATS and subsequently enlisted in the RAF where he went on to have a successful flying career; airmen who had taken the tests in the "guise of recruits" and subsequently failed; the fear that the inaccuracy of tests led to the rejection of those who were suitable. With regard to the last, Biesheuvel pointed out that testing was only one part of a process of recruitment—there was too the interview as well as scholastic and academic performance. However, aptitude testing was about determining "potentiality rather than achievement", it was less about what one had already done than about determining one's capability. Perhaps the greatest concern raised by Biesheuvel—and one for which he had no ready answer—was that aptitude testing could be "taken to its logical conclusions": the total reduction of the individual to a set of

[59] Biesheuval, "The Psychological Care of Combatants".

statistics that would determine every facet of their life with the complete loss of human agency.[60] It was a reminder of his work in education and the use of intelligence testing where he challenged segregationist thinking that limited opportunity based on race.

1942 was a watershed year during the War with the fall of Tobruk and a shortfall in recruiting. The year also marked the implementation of a policy of dilution in the UDF where the black non-combatants of the Non-European Army Services (NEAS) were attached to previously all-white units, thereby freeing the latter for combat. Black men sans military experience, with a lack of training and education, were now expected to work alongside their more experienced white counterparts.[61] It was also in January 1942 in a speech to the Institute of Race Relations that Smuts uttered the oft-quoted, "…segregation has fallen on evil days…" His words were in part an acknowledgement of the massive industrialization initiated by the War that had contributed to a burgeoning urbanized, black working class. As Hyslop argues, while Smuts's words were ultimately unfulfilled rhetoric, they nevertheless signalled a changing attitude towards an urbanized black population, allowing for greater intervention on the part of welfare organizations to address their social and economic needs and an acknowledgement of government responsibility.[62] It was within this context then that Biesheuvel would find his post-war calling.

The Council for Scientific and Industrial Research (CSIR) was formed in 1945 under Basil Schonland. One of its specialist departments was a "national bureau of personnel selection". Lieutenant-Colonel Simon Biesheuvel whose experience in the South African Air Force Aptitude Test Section during the Second World War made him the ideal candidate to head this new department that would become the National Institute of Personnel Research. Biesheuvel was released from military service on 3 April 1946, then to take up his new posting. The experience of aptitude testing and selection of SAAF personnel was now to be applied to

[60] Biesheuvel, "Scientific Personnel Selection".

[61] Ian Gleeson, *The Unknown Force: Black, Indian and Coloured Soldiers Through Two World Wars* (Rivonia, Ashanti Publishing, 1994), pp138–139.

[62] Jonathan Hyslop, "'Segregation has fallen on evil days': Smuts' South Africa, Global War, and Transnational Politics, 1939–46", *Journal of Global History*, Vol 7, 438–460, 2012, p452.

industry and to education.[63] Aptitude testing, whether applied to white pilots during a world war or black and white workers in a post-war world would be an important component in creating a modern South Africa.

[63] "New Bureau will Provide Industry Aptitude Tests", *Rand Daily Mail*, 14 January 1946; Dubow, "Mental Testing", p163 and "Biesheuvel, Simon", Service Card No W.R.12/7542, 203333V, DOD.

CHAPTER 9

"The Amoebic Beginning of a World Mind…:" Sharing Knowledge in a Post-War World

Speaking at the Conference on Science and World Order in 1941, science fiction writer Herbert George Wells claimed that there existed currently a state of "world dementia", of scattered, unco-ordinated and random impulses that provided little way forward for the world. He proposed that it was the role of scientists to formulate a relationship between "organized will and knowledge" and create a blueprint for a world order based on "greater humanity, community and solidarity". As reported in *Nature*, "Science is man's consciousness". A first step in realizing this "world mind" was the creation of a "world language", a means of communication that was shared among all people, where meanings would not change over time and were clear-cut so as to prevent misunderstanding. At the core of this concept of the "world mind" was the creation of a repository for all the world's knowledge that could be accessed readily by every person across the globe, a collation of "the world's thinking and learning", continuously updated and shared among all.[1] Occurring within the

[1] "Science and the World Mind", *Nature*, 11 October 1941, Vol 148, p426.

"The Amoebic Beginning of a World Mind…": Title taken from "Conference on Science and World Order", *Nature*, 4 October 1941, Vol 148, p391.

S. Chetty, *Science and Technology in South Africa, 1939-1946*, Palgrave Studies in the History of Science and Technology,
https://doi.org/10.1007/978-3-032-05084-7_9

context of a war marked by national and imperial ambition, it is not difficult to empathize with Wells's view of a world united in common purpose, with knowledge key to progress and the attainment of a utopia.

Wells's utopian global vision was by no means new, with a similar sense of purpose and perhaps of euphoria evident in the post-First World War period. American president Woodrow Wilson's Fourteen Points promoted a vision of democracy and of freedom in the wake of the carnage of the First World War where the existing "system of [the] balance of powers" had failed. In his speech, made in January 1914, Wilson articulated his vision for fair trade, a reduction in armaments, the notion of self-government for all states and, significantly, a "League of Nations" that would replace the old system and ensure that global peace and order were maintained. For Wilson, freedom and democracy were universal, they were "the principles and policies of forward looking men and women everywhere, of every modern nation, and of every enlightened community". And his vision would resonate with the marginalized and subject peoples all over the world.[2]

While the League of Nations had initially not paid particular heed to "intellectual" aims, a report by the 2nd Committee of delegates to the Assembly of the League of Nations, emphasized the importance of "intellectual work".[3] The "Report on the Organisation of Intellectual Labour" drew attention to the work of the Union of International Associations which had been in existence since the late nineteenth century. The aim of the Union of International Associations was remarkably similar to Wells's vision of a world mind: "The principle aim of the promoters of this Union was to give the world a centre for the accumulation, co-ordination and diffusion of all that the human race has thought and realized in the intellectual sphere. Its ambition is to give the human race a memory and a brain". This was done through the creation of a record of publications (some twelve million in 1920), a "Documentary Encyclopedia" holding more than a million documents, an "International Museum" and a "Collective International Library". At the heart of the idealism of the Union of International Associations was globalization and the intellectual co-operation "of the thinkers of the world". In this spirit, the Report

[2] Margaret Macmillan, *Paris 1919: Six Months that Changed the World* (London, John Murray Publishers, 2019), pp21–23.

[3] "League of Nations: international Cooperation", United Nations Library and Archives, https://libraryresources.unog.ch/lonintellectualcooperation/ICIC, accessed 5 May 2025.

recommended the formation of an organization affiliated to the League of Nations that would foster the work of the Union and to "[develop] a liberal spirit of good-will and world-wide co-operation".[4] In September of the following year, the League of Nations adopted Resolution 4 to create a committee to address "international co-operation" that would comprise men and women. The resolution was based on the realization "that the future of the League of Nations depends upon the formation of a universal conscience" with the international dissemination of "the ideas which can ensure peace among the peoples".[5] The result was the International Committee on Intellectual Cooperation that included among its members Marie Curie and Albert Einstein. Yet, despite the utopian ideal of global intellectual collaboration, there remained division and difference. Prior to a meeting of the Committee in August 1922, Einstein decided not to attend due to anti-Semitism in Germany, making him the less-than-ideal candidate "to serve as a liaison between the German intelligentsia and the international intelligentsia", an ominous harbinger of what was to come. Further, Germany was also still not yet a member of the League of Nations, indicative of the divisions sowed by the recent War—and would only join four years later.[6]

The year of Wells's speech also saw the signing of the Atlantic Charter which, like Wilson's Fourteen Points, envisioned a post-war world marked by democracy and equality. Yet war only highlighted the tension between intellectual co-operation and division. In contrast to Wells's advocacy of the free exchange and availability of knowledge, arguably the greatest attempt at secrecy associated with state and ideological security among the Allies was the Manhattan Project. While ostensibly fighting a war against the totalitarian threat presented by Germany and Japan, western

[4] "Annex E: Report on the Organisation of Intellectual Labour" presented by M Lafontaine, 18 December 1920, pp769–771, United Nations Library and Archives, https://libraryresources.unog.ch/lonintellectualcooperation/ICIC, accessed 5 May 2025.

[5] League of Nations, "International Organisation of Intellectual Work", 21 September 1921, United Nations Library and Archives, https://libraryresources.unog.ch/lonintellectualcooperation/ICIC, accessed 5 May 2025.

[6] Dava Sobel, *The Elements of Marie Curie: How the Glow of Radium Lit a Path for Women in Science* (London, 4th Estate, 2024), pp184, 196–197.

countries were anything but oblivious to the authoritarian and Communist Soviet Union. Stalin's apparent exclusion[7] from knowledge about the development of the atomic bomb was a harbinger of the suspicion and hostility that would mark the relations between the West and the Soviet Union and usher in the Cold War, destined to continue for almost five decades. However, by 1946, the "father of the atomic bomb", J. Robert Oppenheimer was articulating his desire for the creation of a "world government" that would control and regulate the use of atomic energy, ensuring its benign rather than belligerent use. Drawing upon Niels Bohr's notion of "openness" in the sharing of scientific and technological knowledge, it was evident that so powerful an invention would irrevocably change notions of "national sovereignty", compelling global co-operation. While acknowledging that a "world government" was not entirely feasible at that point, Oppenheimer nevertheless believed that total "transparency" was necessary to prevent the further manufacture—and ultimate use—of atomic weapons. It was only through international co-operation that scientific and technological knowledge could be used to maintain peace.[8] And co-operation and communication were integral to the formation of scientific organizations and the holding of scientific conferences with both epitomizing a complex interplay between the national, the regional and the global.

As Bigg et al. discuss, while conferences were increasingly common from the nineteenth century, they are understudied as a site of analysis for their (often informal) scientific communication that have been as integral to the dissemination of knowledge as the scientific journal. Further, conferences function as a means of forming a scientific community which, in the case of international conferences, transcends national boundaries. In the twentieth century, the scientific conference has also been enmeshed in the global political order where aspirations to "internationalism" have been tempered by ideology and shifting power relations.[9] The co-ordination and communication of scientific knowledge which

[7] Soviet spies were present at Los Alamos and the Soviet Union was therefore aware of the development of the bomb.

[8] Kai Bird and Martin J. Sherwin, *American Prometheus: The Triumph and Tragedy of J. Robert Oppenheimer* (London, Atlantic Books, 2009), pp288–289, 342.

[9] Refer to Charlotte Bigg, Jessica Reinisch, Geert Somsen and Sven Widmalm, "The Art of Gathering: Histories of International Scientific Conferences", *The British Journal for the History of Science*, 56, 2023, pp423–433.

form the focus of this chapter is therefore contextualized by a global war, embryonic nationalism, a waning empire and a seismic shift in the balance of power that would once more polarize the world. Yet, while built on prevailing inequalities and difference, there existed too an optimistic and almost euphoric vision of a new global order based on international co-operation that would surmount the nationalist conflict that had characterized the War. Science would be harnessed to development and to progress; it would bring this new order into being.

Creating a National Organization

The Second World War officially ended with the surrender of Japan after the atomic bombing of Hiroshima and Nagasaki in August 1945. Five months later, a meeting of the Standing Committee of the Heads of Departments took place at the Council for Scientific and Industrial Research (CSIR) in the South African administrative capital, Pretoria. The departments represented included Agriculture, Forestry, Posts and Telegraphs, Lands and Defence. The various figures responsible for South African industry, agriculture, communications, infrastructure and defence had assembled to discuss matters pertaining to the forthcoming Commonwealth Scientific Conference to take place in Britain in June of that year.[10]

The chair of the meeting, representing the Prime Minister's Department, was Dr B.F.J. Schonland, who would be part of the delegation representing South Africa at the conference, and was also the first president of the CSIR. Formed in 1945, the CSIR was an emulation of similar institutions across the Commonwealth.

Sir Richard Gregory, former editor of *Nature* and president of the British Association, was an ardent proponent for the role of science in society as instrument of progress. He was also involved in establishing the Division for the Social and International Relations of Science which hosted the Conference on Science and World Order.[11] It was on a visit to India in 1933 that he suggested the formation of an organization dedicated to scientific research and industry, akin to the British Department of

[10] "Minutes of the Meeting of the Standing Committee of Heads of Departments", 25 January 1946, File Q/T 876/1: British Commonwealth Committee Meetings on Post-war Publication of Scientific War Work, KMG, Group 27, Box 138, DOD.

[11] "Sir Richard Gregory, Bart., F.R.S", *Nature*, Vol 153, 1944, p133.

Scientific and Industrial Research formed during the First World War. For Gregory, Indians had an aptitude for science and technology and could apply their knowledge so as to best utilize the "resources of the country" for development. There was some resistance, and funding was only provided to finance a more modest bureau for research in 1935. With the Second World War and the prioritization of the use of resources to aid the war effort—uniforms, dyes, fuel—funding was greatly increased, and the Council for Scientific and Industrial Research was established in September 1942. When India gained independence five years later, the CSIR was headed by the country's first Prime Minister, Jawaharlal Nehru, and harnessed to the modernizing project.[12] The formation of the Indian CSIR, then, was based on the limitations under colonialism, the exigencies of war and the modernizing aspirations of the independent state.

The Australian CSIR was also related to Australian nationalism and the achievement of independence. After the establishment of the Federation of Australia at the very outset of the twentieth century, a decision was taken to "[promote] Australian industry", leading to the formation of the Advisory Council of Science and Industry in 1916. A decade later, the better funded CSIR was established. Its focus was on the development of Australia in the areas of agriculture, mining and industry. With the outbreak of the Second World War, a defence component was added to what was initially envisaged as a civilian agency, and the CSIR was also involved in the development of radar. By 1949, it ended its military ties and became CSIRO (the Commonwealth Scientific and Industrial Research Organization).[13]

1916 also saw the formation of the Honorary Council for Scientific and Industrial Research, later to become the National Research Council of Canada (NRC). Like its Australian counterpart, it was forged in the crucible of war. Drawing upon the inspiration of both the British and Australian organizations, it was the result too of the petitioning of Canadian scientific and business interests for the fostering of "national science". With four divisions encompassing Biology, Mechanical Engineering, Chemistry and Physics, the NRC was poised to mobilize its

[12] Vigyan Prasar, "Shanti Swarup Bhatnagar", http://www.vigyanprasar.gov.in/scientists/ssbhatnagar/ShantiSwarupBhatnagar.htm, accessed 24 April 2024.

[13] Colin Ward, "Brief History", 10 December 2015, *CSIROpedia*, https://csiropedia.csiro.au/our-history/, accessed 24 April 2024.

resources for the Second World War. Its war research encompassed food, gases, aircraft, radar, the infamous iceberg carrier (as discussed in Chapter Two) and the atomic bomb. National aims notwithstanding, the war work of the NRC was intricately linked to research carried out in Britain and in the United States.[14]

While the South African CSIR was formed late in relation to the British and other Commonwealth research bodies, its genesis lay in a report prepared by H.J. van der Bijl in 1921 who made a persuasive case for the importance of government involvement in advancing and funding research into industrial development. Prescient of the aims of the CSIR two decades later, Van der Bijl stressed the use of natural resources in underpinning industrial development, recommending the establishment of an organization for just such a purpose. As Director General of War Supplies during the Second World War, Van der Bijl was essential in co-ordinating industry to meet the demands of war, in conjunction with the other Dominions. The War spurred the growth of manufacturing industry and transformed the South African economy. It also brought a growing awareness of the importance of collaboration in scientific and technological development as experts co-operated in finding solutions to the challenges brought by war on a hitherto unparalleled industrial scale. In early 1945, Jan Smuts assigned B.F.J. Schonland the post of Scientific Advisor to the Prime Minister. In his new role, Schonland drew up a Bill that proposed the establishment of an institution for the development of research that would culminate in the CSIR. In developing his proposal, he drew upon his existing experience with various wartime organizations and was inspired in particular by the set-up of similar institutions in Australia and Canada. Introducing the Bill in Parliament in May 1945, Acting Prime Minister Jan Hofmeyr spoke of South African scientific achievement but identified a significant shortcoming: a failure "in the general organization of our research activities and their co-ordination..." It was a failure that the CSIR was designed to remedy. The Bill was passed to unanimous approval in a country at a time when political divisions were increasingly insurmountable, demonstrating the importance accorded to

[14] Refer to Donald J.C. Phillipson, "The National Research Council of Canada: Its Historiography, its Chronology, its Bibliography", *Scientia Canadensis*, Vol 15, No 2(41), 1991, pp177–193.

scientific and technological development by both the Smuts government and its Afrikaner nationalist successors from 1948.[15]

B.F.J. Schonland was named chair of the CSIR. Other council members included H.J. Van Eck, R.D. Wilcocks, Bernard Price and S.H. Haughton.[16] As head of the Industrial Development Corporation, Van Eck had played a significant role in driving South African wartime industrialization.[17] Wilcocks had been a member of the Carnegie Commission in depression-era South Africa, tasked with investigating the "poor white problem" and had a special interest in intelligence and aptitude. Bernard Price—who was to die less than three years later—was a British immigrant who had worked as an electrical engineer in South Africa in 1909 when he was appointed chief engineer for the Victoria Falls and Transvaal Power Company, which was responsible for generating electricity for the country's industrial heart, the Witwatersrand. He was also head of the Advisory Committee on Science and Technology during the First World War.[18] Sidney Haughton was a geologist, awarded his degree from Cambridge University who, in his capacity as director of the Geological Survey, served on bodies dedicated to "strategic resources" during the War such as the Fuel Research Institute and the Industrial and Agricultural Requirements Commission.[19]

In October 1945, Jan Smuts opened the inaugural CSIR meeting at the Union Buildings, the seat of government. He expounded on the aims of the Council in his address: its role in enabling science and industry similar to that of organizations in other parts of the Commonwealth; to provide funding to research and tertiary institutions that would further develop industry in South Africa; and to advise the government on how best to utilize South Africa's "natural resources". In his response to the Prime Minister, Schonland saw the CSIR as a continuation of the war effort but in the context of a post-war world, his language evocative of

[15] D.G. Kingwill, *The CSIR: The first 40 years* (Pretoria, CSIR, 1990), pp5–10.

[16] "Smuts Calls for Great Industrial Research Move", *Rand Daily Mail*, 9 October 1945.

[17] Refer to Chapter Two on civilian invention.

[18] "Price, Mr Bernard", *S2A3 Biographical Database of Southern African Science*, https://www.s2a3.org.za/bio/Biograph_final.php?serial=2247, accessed 27 April 2024.

[19] "Haughton, Dr Sidney Henry", *S2A3 Biographical Database of Southern African Science*, https://www.s2a3.org.za/bio/Biograph_final.php?serial=1238, accessed 27 April 2024.

the military ethos which had dominated public life for the preceding six years and of his own role in the War, "We have a contribution to make to the world's stock of knowledge. You have given us our operation orders. This is our D-Day...".[20]

Science and the World Order

Basil Ferdinand Jamieson Schonland was born in Grahamstown in the Eastern Cape in 1896, a frontier region in South Africa. His father, Selmar, was a German-Jewish scientist with a passion for botany. Schonland was awarded his BA degree from the University of the Cape of Good Hope and had enrolled for the Mathematical Tripos at Cambridge University before receiving a commission as Second Lieutenant in the Royal Corps of Engineers during the First World War. As part of the Signal Section, he served as a wireless operator during the Battle of the Somme, subsequently training junior officers in the use of what was still a relatively new technology. With the outbreak of the Second World War, Schonland's wireless expertise was called upon in the development of radar and he was subsequently attached to the Army Operational Research Group based at Richmond Park in London. In 1944, he was appointed Scientific Advisor to General Bernard Montgomery and tasked with finding scientific solutions to potential problems that could arise with the planned D-Day invasion. These included determining the impact of Allied bombing as well as the bizarre possibility of the Germans "electrifying" the sea so as to thwart the potential landing. Schonland retained however, an interest in the development of science in his home country, and was especially energized by discussions around the creation of research institutes in Africa. As a result, he presented a talk to a committee of the Royal Society, presided over by Sir William Bragg, entitled "Lost Chances and Science in Africa".[21]

Unsurprisingly then, Schonland was one of the candidates considered for an international conference held during the height of the War in late September 1941. The Conference on Science and World Order was held in the wake of the Atlantic Charter with its vision for a free and equitable

[20] "Smuts Calls for Great Industrial Research Move".

[21] Brian Austin, *Schonland: Scientist and Soldier* (Bristol and Philadelphia, Institute of Physics Publishing, 2001), pp1–2, 19, 22–23, 46–48, 170–171, 242, 246–247, 271–279.

post-war world that had been proposed by American President Franklin Roosevelt and British Prime Minister Churchill in August of that year. 1941 had been a challenging year for the Allies with the Blitz, setbacks in Greece and Tobruk and the German invasion of the Soviet Union. While the United States had yet to officially enter the War, Roosevelt signed the Lend Lease Act to provide weapons to Allied forces. In the midst of what would become a protracted conflict, the Charter then was an idealistic vision articulating "their hopes for a better future of the world". One of the principles of the Charter that would provide hope and impetus to the colonized people of Africa and Asia was that of self-determination. A key tenet of the Charter was the notion of equality—equality in trade and access to resources, equality in terms of quality of life and "freedom from fear and want".[22] Amidst what seemed like perpetual strife and uncertainty, it was a utopian vision for a post-war world.

For South Africans, the Science and World Order Conference was to be of special interest to Africa in its desire to "give reality" to the principles of the Atlantic Charter with Lord Hailey expected to give serious consideration to science as it related to the African context and "the problems of famine, disease and soil rehabilitation".[23] The Conference was organized by the Division for the Social and International Relations of Science, a relatively new sector within the British Association for the Advancement of Science (BAAS). It was held over three days in late September at the Royal Institution in London, comprised six sessions with presentations given by delegates from as far afield as the United States, China and the Soviet Union. In the case of the last, however, the Soviet Union was represented by its ambassador to Britain, M. Maisky. The German invasion had prevented their attendance and German "jamming" precluded them from sending their speeches via radio—as South Africa's Jan Smuts would do. Soviet scientists nonetheless "expressed their solidarity" with other Allied scientists in the achieving of victory while Maisky cautioned against the expectation that much could be achieved in terms of "world

[22] "The Atlantic Charter: Declaration of Principles issued by the President of the United States and the Prime Minister of the United Kingdom", 14 August 1941, *North Atlantic Treaty Organization*, https://www.nato.int/cps/en/natohq/official_texts_16912.htm, accessed 23 April 2024.

[23] "S.A. Scientists Wanted for Big Conference: British Association to Discuss African Problems", *Rand Daily Mail*, 9 September 1941.

planning" over a short conference, in contrast to the decades-long system in place in the Soviet Union.[24]

From the outset, as evident in a luncheon address by the Secretary for Foreign Affairs (and later Prime Minister), Anthony Eden, science was to be harnessed to "public affairs", and the themes of the conference related to the role of science in government and society—although concerns were raised regarding the possible adverse effects on perceived scientific neutrality when used for political purposes. An important component throughout, however, was the use of science in creating an equitable world: the removal of "both hunger and the sword" to compel labour and "a scientifically defined welfare standard" that would reduce the precarious and unpredictable nature of the impact of economic forces on the individual. For Hailey in his address on Africa, the new concern with welfare would only be "beneficial" to "colonial policy". Amidst this model vision, the Czechoslovakian president uttered a word of caution: while the conference was about the free use of science to create a better post-war world, it was also about the prevention of "the misuse of great inventions and all kinds of technological progress for criminal and destructive purposes".[25]

While Schonland had been unavailable to attend the conference, the face of South African science and politics, Jan Smuts, recorded a speech which was played to conference attendees on a gramophone. In it, Smuts articulated his own vision for the role of science as a beacon in the midst of the current conflict and of playing an integral role in the creation of a post-war world: "With our victory, science will not merely be reinstated to her honoured status, but a new era will open for her. Our aim is not only more knowledge and ever-new discovery of truth, but also the promotion of social welfare and the building of a great society of free people".[26] Smuts would contribute to the creation of this new world order when he wrote the Preamble to the Charter of the United Nations.

In addition to the influence of the Atlantic Charter, the Science and World Order Conference was also contextualized by the BAAS's policies regarding the relationship between science and society—a relationship

[24] "Conference on Science and World Order", p389.

[25] "Conference on Science and World Order", pp388–390.

[26] "Conference on Science and World Order", p389.

that was partly inspired by their American counterparts. In his dissertation, Peter Collins traces the inspiration for the Conference to a decision taken by the American Association for the Advancement of Science in Philadelphia in December 1940 that stressed the "responsibility of scientific men for the general welfare of free peoples" and called upon the British Association to join them in formulating "an international charter of democracy", one that was to be based on "scientific principles". The response by the British Association was the "Charter of Scientific Fellowship" that in stirring language upheld the values of democracy and intellectual freedom. Scientific thought was seen as the prerogative of all regardless of "[race and class]...with the whole world as its outlook and service to mankind its highest aim". It was an ardent declaration of internationalism to counter the rampant nationalism of war and by July 1941, the Science and World Order Conference had been conceived as a means to determine the ways in which scientific knowledge could be applied to society and the creation of a "new international order".[27]

In 1942, Penguin published "Science and World Order" that articulated the views of the more liberal and leftist-learning scientists that re-iterated the spirit of the Conference. It highlighted the essential role of science and scientists in alliance with government to plan for a better world. Science would be harnessed to society. This image in the midst of war, of a planned and equitable society, nevertheless jarred with the use of science in the service of death and destruction on an unprecedented scale, the work of what pacifist Vera Brittain described as "perverted genius".[28]

A few days after the Science and World Order Conference, a more modest gathering took place at the Royal Society on 7 October 1941. Those assembled considered the creation of a committee comprising representatives from Britain, the Dominions and India that would address both scientific research and the ways in which science could be used to better "the life of the peoples of the Empire" through confronting "technical, biological, medical and economic" challenges in the post-war world. The United States would later send representatives to subsequent meetings as well. Inspired by the collaboration that was increasingly

[27] Peter Michael Digby Collins, "The British Association for the Advancement of Science and Public Attitudes to Science, 1919–1945" (University of Leeds, unpublished PhD dissertation, 1978), pp217–219, 223.

[28] David Edgerton, *Britain's War Machine: Weapons, Resources and Experts in the Second World War* (Penguin Books, 2012), pp141, 150.

becoming a hallmark of the Second World War, the proposal was the creation of a British Commonwealth Scientific Collaboration Committee where all member nations would collaborate on and disseminate scientific research. Its hub was to be London where the Dominions and India were to "[maintain] permanent scientific and technological representation", although a suggestion was to have similar representation "in other capital cities of the English-speaking world".[29] It is in the light of this spirit of continued co-operation and the imagining of a post-War world that the Commonwealth Scientific Conference took place in London in June 1946. It was also—as Macleod suggests—a means of re-envisioning the Commonwealth and Britain's changing relationship with the Dominions and colonies.[30]

The Post-War World

In January 1945, the Royal Society made an early call for a conference that was initially planned to take place after victory had been achieved in Europe. The aim of the conference was to review the current challenges facing the Commonwealth and work towards collaboration in resolving them. The conference was to be a gathering of the keenest scientific minds in the Commonwealth: "We are most anxious that the conference should be a conference of scientific men, whose advice carries weight in their respective fields and to whose counsel authority is accorded in their several countries because of their scientific achievements and experience".[31]

To prepare for what would become the unofficial Royal Society conference as well as the Commonwealth Scientific Conference, a panel was set up comprising representatives from the Commonwealth and from India. Major Eric Boden was the Scientific Liaison Officer representing South

[29] "The Activities of the Society", *Notes and Records of the Royal Society of London*, Vol 4, No 1, April 1946, pp6–7.

[30] Roy Macleod, "The Royal Society and the Commonwealth: Old Friendships, New Frontiers", *Notes and Records of the Royal Society of London*, Vol 64, Supplement 1: The Royal Society and Science in 20th Century, 137–149, 2010, pp140–141.

[31] Secretaries, Royal Society, 22 January 1945, C&O 85/46—Post-War Scientific Collaboration: Official Civil Science Panel and the Commonwealth Scientific Conference Working Party, IOR/L/E/8/4759, British Library (hereafter BL).

Africa and the CSIR.[32] At the outbreak of the War, Schonland had been tasked with developing a system of radar that was to be used to protect the important supply route around the Cape. He assembled a task team that would eventually become the Special Signal Services (SSS). Boden joined the SSS and was eventually part of a team that established a radar station on Signal Hill in Cape Town.[33] As Scientific Liaison Officer based in London at the end of the War, Boden served in Schonland's stead when the need arose.

Prior to their attendance of the Conference (that would be held a year later than initially planned), the South African delegation had to consider the extent to which scientific knowledge could be shared with the Commonwealth. Although the CSIR had been established relatively late in contrast to other research institutions, it was part of a civilian network of knowledge-sharing across the Commonwealth. All research falling under Britain's Department of Scientific and Industrial Research (DSIR) was usually published, with reports sent to the equivalent research institutes in the Dominions. In turn, the DSIR received copies of all publications issued by the Dominions' various research institutes as well as that from India.[34] Yet there was caution regarding the sharing of potentially sensitive military technology which overrode Commonwealth collaboration.

With the conclusion of the War, censorship was no longer in place however South African military personnel—both those who had had left the service as well as those still on active duty—were governed by the Official Secrets Act—which related to civilians divulging secret information—and Orders and Instructions 20 and 21. In the case of the Official Secrets Act, military personnel (some of whom were now beginning to write down their experiences of war) were encouraged to submit their manuscripts to the UDF to determine if they contravened the Official

[32] "Commonwealth Scientific Official Conference, July 1946: statements and memoranda prepared by the Working Party appointed by the Civil Science Panel", January 1946, CO 1073/321, TNA.

[33] Geoffrey Mangin and Sheilah Lloyd, "The Special Signal Services (SSS): We scanned the seas and skies in the Second World War", *Military History Journal*, Vol 11, No 2, 1998, http://samilitaryhistory.org/vol112ml.html, accessed 7 June 2024.

[34] "Commonwealth Scientific Official Conference, July 1946".

Secrets Act. However, any contravention of the Act was a civilian case and would be subject to civil action.[35]

Orders and Instructions 20 and 21, however, were directed specifically at military personnel and, at its heart, it was unsurprisingly related to the preservation of state security:

> An officer or soldier is forbidden to communicate any military information which might directly or indirectly assist an enemy, to any person other than a person to whom he is authorized to communicate it, or a person to whom it is, in the interests of the State, his duty to communicate it.[36]

Further, military personnel were forbidden from communicating information or their views on military matters to friends, family and the press or from publicly questioning the methods and instructions from their superior officers. Any "professional" information that they acquired while in the service of the UDF, remained the property of the UDF. Should the serviceman want to publish an article pertaining to their military service, approval first had to be sought from the Chief of the General Staff via the commanding officer.[37]

The Committee on Postwar Publication of Scientific War Work was established at the end of the War with its mandate "to prevent the publication of any Scientific War Work" that was still considered classified. In a meeting of the Committee in March 1946, Schonland requested the policy as applied to military personnel viz. Orders and Instructions 20 and 21 and made specific inquiry as to what extent the work done by UDF personnel on gases and mine detectors was to remain a secret.[38] With regard to gases, he was informed that all information related to the development of such was to remain classified however

[35] G.N. Nauhaus, "Committee on Post War Publication of Scientific War Work", 8 March 1946, File Q/T 876/1: British Commonwealth Committee Meetings on Post-war Publication of Scientific War Work, KMG, Group 27, Box 138, DOD.

[36] "Communication of Military Information", File Q/T 876/1: British Commonwealth Committee Meetings on Post-war Publication of Scientific War Work, KMG, Group 27, Box 138, DOD.

[37] "Communication of Military Information".

[38] Major J.H. Pienaar, "Committee of Postwar Publication of Scientific War Work", 4 March 1946, File Q/T 876/1: British Commonwealth Committee Meetings on Post-war Publication of Scientific War Work, KMG, Group 27, Box 138, DOD.

there were no restrictions in place with regard to the divulging of knowledge regarding mine detectors.[39] This, then, would give Schonland the clarity required regarding the dissemination of scientific knowledge at the Commonwealth Scientific Conference.

But consideration had to be taken too of the patent rights of individuals. While the UDF acknowledged the "national importance" of the Commonwealth Conference, the "purely" rather than "[applied] scientific aspect" of the Conference meant that it held little appeal for a defence force focused on the technology of war. However, while the UDF did not plan to send a representative to the Conference, they nevertheless submitted to Schonland a proposal for a Commonwealth Patent which the latter was to present at the conference following that held by the Royal Society.[40] In contrast to David Edgerton's view of patents having little utility and being an inadequate reflection of technology due to their haphazard implementation,[41] for the UDF, Patent Offices were the repository of the current technological knowledge of South Africa and the rest of the Commonwealth. The belief was that all "technical research" is inevitably reflected in patent applications and the constant influx of applications from South Africa and all over the world reflected the latest in technological innovation and invention, both actual and theoretical. This vast wealth of untapped technological potential had been simply "pigeon holed" and thus neglected. The proposal made by the

[39] Nauhaus, "Committee on Post War Publication of Scientific War Work". During the interwar period, South African Minister of Defence Oswald Pirow considered the use of chemical weapons against a potentially insurgent black population. There was a focus on the production of mustard gas within the Union, with the hope of British assistance. The UDF's first unit dedicated to chemical warfare was established with the outbreak of the Second World War and, despite their initial reluctance on divulging chemical weapons technology to the South Africans, the British were subsequently involved in the establishment of two chemical weapons factories in South Africa for the production of mustard gas. By 1943, American scientists also visited sites with tentative plans to have mustard gas exported to the United States. By the end of the war, however, Britain withdrew its support of the production of mustard gas in South Africa. Please refer to Tilman Dedering, "South Africa's Secret Chemical Weapons Project, 1933–1945", *The International History Review*, Vol 44, No 2, 263–281, 2022.

[40] Director of Explosives, "Commonwealth Scientific Conference", 29 January 1946, File Q/T 876/1: British Commonwealth Committee Meetings on Post-war Publication of Scientific War Work, KMG, Group 27, Box 138, DOD.

[41] David Edgerton, *The Shock of the Old: Technology and Global History Since 1900* (London, Profile Books, 2008), p200. This is addressed in Chapter Two.

UDF then was for a central hub—most likely to be located in Britain which had the resources to investigate the feasibility of the patent—where patents were filed and records of these would also be made available to other patent offices throughout the Commonwealth. At the same time, however, patents addressing specifically and uniquely local or national needs would still be filed at local patent offices—however the Commonwealth Patent would "supercede" the local patent. In a manner not unlike H.G. Wells's proposal of a "world mind", the centralization of patent information would then make easier the co-ordination of and access to technological information throughout the Commonwealth. The availability of the information would allow for the researcher to be dispatched to where their skills were most required; the co-ordination of research across the Commonwealth as well as the resources required to fund the development of particular inventions; and the growing status of the proposed central patent hub would attract more applicants from within the Commonwealth but also from without, thereby giving "early insight into any technical enterprise". And in the spirit of the envisaged post-war world, it would fulfil two functions: it would record and make known all inventions, regardless of their importance, and it would create a better world for the people of the Commonwealth through the judicious development of technology that would "stimulate manufacturing and trade".[42] The Commonwealth Patent as suggested by the UDF addressed a similar concern raised by the British. While the sharing of information was an imperative, it served little purpose until it could be translated into practical action. Another proposal to be put forward for discussion at the Conference was the creation of local organizations responsible for "patent exploitation and administration". Any "useful discovery" should be developed and either be put "into production" or have their specifications made available "throughout the Commonwealth".[43]

In addition to the Commonwealth Patent, the UDF also requested that the South African delegation give consideration to specifically medical topics related to the war at the Conference: the field of aviation medicine, "army public health" with an emphasis on work done on malaria and

[42] G.N. Nauhaus, "Commonwealth Scientific Conference", 26 January 1946, File Q/T 876/1: British Commonwealth Committee Meetings on Post-war Publication of Scientific War Work, KMG, Group 27, Box 138, DOD.

[43] "Commonwealth Scientific Official Conference, July 1946".

"army rations", as inspired by the experiences of prisoners of war.[44] Stimulated by the recent conflict, UDF suggestions were also made in light of the proposed fields of interest for the conference that had been forwarded to the South Africans.

The conference was to take place in two parts: the first was the Royal Society Empire Scientific Conference from 17 June to 8 July 1946, which was a precursor to the "Official Conference". While the latter dealt with co-operation and the sharing of scientific and technological knowledge throughout the Commonwealth, the former was to focus on the key challenges related to science and address scientific "progress" over the course of the War. However, also on the agenda, was a deliberation regarding the "closer collaboration" between scientists as well as scientific institutions across the Dominions—and India. To achieve the latter, the Society would possibly make recommendations regarding areas of research that needed further attention. The input of the Royal Society Conference could then be considered at the Official Conference before it made its own recommendations.[45]

As evident in the preliminary programme sent to the South Africans, the Royal Society Conference would take place over approximately three weeks with activities planned for delegates throughout. The mornings would be allocated for discussion, and this was to be followed by site visits to "laboratories and exhibitions" with talks held on particular areas of research that would appeal to different sectors of attendees. The daily activities would conclude with more "informal discussions"—also likely to be specialized—at Cambridge and Oxford. At the outset of the Conference, delegates were requested to submit a report regarding the state of scientific organization in their respective countries: the organization of institutions, the role of academic institutions and of industry, as well as "their chief problems in applied science". Should the member countries be agreeable then, Britain would make the first presentation, followed by New Zealand, Canada and the "Colonial Empire", with India, South Africa and Australia on the third day.[46]

[44] Nauhaus, "Commonwealth Scientific Conference".

[45] "Royal Society Empire Scientific Conference: Outline of Programme", File Q/T 876/1: British Commonwealth Committee Meetings on Post-war Publication of Scientific War Work, KMG, Group 27, Box 138, DOD.

[46] "Royal Society Empire Scientific Conference".

Following the reports, prospective delegates were advised on the scientific topics that the Conference was to address—and on which they were then expected to contribute. These included medicine, with Britain expecting to present on the "physical and physiological" effects of working "under tropical conditions and in industry" as well as the transmission of disease, especially those transmitted by insects.[47] The concern regarding the impact of tropical diseases on military forces had a long history in the British armed forces. In the twentieth century, the First World War had seen outbreaks of disease such as malaria during the East African campaign and in the Middle East, with the Royal Army Medical Corps (RAMC) maintaining its research into tropical medicine during the interwar period. During the Second World War, tropical disease took a particular toll on troops in the Far East, although malarial outbreaks were curbed with a better understanding of preventative measures. In the Mediterranean, some 25,000 British troops would be hospitalized as a result of sandfly fever.[48] Related to medicine and health was nutrition with a focus in particular on the nutritional health of the "indigenous peoples" of the colonies. Other general topics included challenges in agriculture and in "land utilization and conservation".[49] While addressing the southern African context in particular, William Beinart discusses the preoccupation of colonial officials and agricultural experts with conservation and agriculture in the first half of the twentieth century. Emphasis was made on the importance of the technical management of and intervention in existing land use so as to conserve and replenish natural resources which had been subject to wanton destruction through unscientific farming practices. This remained a British concern in the last decades of the empire on the African continent and was given impetus by the drought in the United States in the 1930s and related fears regarding soil erosion.[50] The use of aircraft—and radio—was another item proposed for the agenda in terms of mapping and surveying the empire. On a practical level, scientific co-operation was to be evident in a uniform system

[47] "Royal Society Empire Scientific Conference".

[48] Mark S. Bailey, "A brief history of British military experiences with infectious and tropical diseases", *J R Army Med Corps*, 159, 150–157, 2013, p154.

[49] "Royal Society Empire Scientific Conference".

[50] Refer to William Beinart, "Soil Erosion, Conservationism and Ideas about Development: A Southern African Exploration, 1900–1960", *Journal of Southern African Studies*, Vol 11, No 1, 52–83, 1984.

of measurement and agreed-upon symbols, creating a coherent scientific language, that would buttress the co-operation between scientists and scientific organizations as well as the availability and distribution of knowledge to the public and to industry—and to be stored for consultation in libraries and other repositories. There would be too, a consideration of the measures that needed to be taken to ensure the safe transportation of floral and faunal samples across borders for study.[51]

Related to the general discussion were the proposed—and far more specific—topics for the afternoon including "nuclear research"—with no indication if this was for the purposes of weaponry or the generation of energy.[52] For a segment of the South African press however, "the keenest scientific brains in the Commonwealth" were to play a role in helping Britain "catch up" to the United States which had been less than forthcoming in divulging the "secrets" of the bomb. Commonwealth collaboration would thus provide Britain with the intellectual "resources" to retain a position of eminence in developing the bomb.[53] As Edgerton discusses, the development of the atomic bomb was a particularly sore point for the British who had begun their development of the atomic bomb from mid-1941—with some assistance from the Canadians. The Americans decided to pursue their own atomic bomb project later in the year, making use of British research. Eventually, the Americans—whose project proceeded rapidly—were less inclined to share their own research and would make use of British scientists, thereby setting back the British endeavour. With the successful use of the bomb against Japan bringing the War to an end, making the United States the first atomic superpower, the Americans were dismissive of any British contribution.[54] The Conference, then, was portrayed as less about a sharing of knowledge than the means by which Britain could retain a strong position in relation to the United States, even as the balance of power had shifted across the Atlantic. Significantly, Britain could not do this alone but required the contribution of the Dominions and colonies to achieve technological parity with its most powerful ally. As Roy Macleod claims, "The

[51] "Royal Society Empire Scientific Conference".

[52] "Royal Society Empire Scientific Conference".

[53] "Schonland Will Share in Empire Atomic Bomb Talks", *Rand Daily Mail*, 2 March 1946.

[54] Edgerton, *Britain's War Machine*, pp119–120.

Commonwealth might serve the mother country in peace, as it had in war".[55]

One of the possible outcomes of the Conference which was of special interest to the British was the establishment of "a permanent British Commonwealth Scientific Office [BSCO]" to house the various "Scientific Liaison Offices" representing the Commonwealth. Yet there also needed to be a consideration of national interests when it came to science and, although the proposed headquarters of the BCSO was to be in London, collaboration between the liaison offices was to be based on exigency and the prerogative of the relevant liaison officers. The latter was ultimately answerable solely to their respective governments and "[retained] complete liberty of action". The proposal was inspired by the BSCO based in Washington D.C. during the War which had comprised the United Kingdom and South African Scientific Missions as well as the Scientific Liaison Offices of Australia, New Zealand and Canada. It had been a hub for the "exchange of information...[in] applied science" and had played a role in streamlining problem-solving through co-ordination and collaboration.[56] There is the sense here that, in addition to pooling the scientific expertise of the Commonwealth, developments in science and technology as well as post-war challenges necessitated collaboration across national borders.

Other areas of interest proposed for the Conference—that admittedly did not capture the popular imagination in the same vein as the atomic bomb—included geophysics, the use of radio communication, meteorology and climate. Deliberately excluded from the initial programme was aeronautics as special provision was to be made for this. In addition, all suggested topics could be adapted depending on the input provided by potential delegates.[57] It was however, in line with these early guidelines—and the emphasis on co-operation and the sharing of information—that the UDF proposed a Commonwealth Patent that would consolidate, develop and disseminate the inventions made across the Commonwealth—and beyond.

[55] Macleod, "The Royal Society and the Commonwealth", p141.

[56] "Commonwealth Scientific Official Conference, July 1946".

[57] "Royal Society Empire Scientific Conference". An aeronautical conference was to take place before the Commonwealth Conference and it was proposed that cognisance would be taken of the findings at the Commonwealth Conference - "Commonwealth Scientific Official Conference, July 1946".

Schonland was responsible for collating the contributions of the South African contingent. According to the recommendations sent by the Royal Society, each country should be represented by approximately eight delegates. Schonland's suggestion was that this be broken down into two from tertiary institutions, two from the CSIR (including himself), a figure from the social sciences, from medicine, a civilian scientist, representatives of the Departments of Mines and Agriculture, respectively, and a member of the Treasury. The CSIR representatives would also be able to take on two of the other scientific roles as well. He was confident that the broad spectrum would be sufficient to present the challenges facing South Africa in both "pure and applied research". Specialists would only be required at a later stage. For the second "Official Conference", called to address co-operation and collaboration, Schonland recommended a subset of the Royal Society delegation.[58]

For Schonland, it was especially important that the delegates represent the areas in which they "could make a definite contribution" and these included "Agriculture, Medicine, Biology and Geology". The focus would remain "pure" rather than "applied" science and some of the suggestions forwarded to Schonland in relation to specialized research included the development of insecticides, "Virus X" as found in potato seeds, the use of microfilm and the composition and physiological effects of the dust prevalent in mines. Recommendations were also made for the training and exchange of specialists across the Commonwealth. Of particular note—and a proposal which would come to fruition with the African Regional Scientific Conference in 1949—were suggestions for the "easy movement of scientists among the various research institutions in Africa" as well as the holding of regular "Pan-African conferences".[59]

A further recommendation made by the British was that the heads of the various national delegations form a "Steering Committee" which would in turn establish a "Drafting Committee". The latter would be expected to liaise each evening and draw up proposals to be put to the conference the following day. Both committees were to determine how to put into effect the recommendations made during the Royal

[58] B.J. Schonland, "Commonwealth Scientific Conference", 17 November 1945, File Q/T 876/1: British Commonwealth Committee Meetings on Post-war Publication of Scientific War Work, KMG, Group 27, Box 138, DOD.

[59] "Minutes of the Meeting of the Standing Committee of Heads of Departments".

Society Conference.[60] It was clearly envisaged as a consultative process throughout—and one which was expected to have real and practical results.

Schonland led the delegation dispatched to Britain in June of 1946. Other members of the South African contingent included P.J. du Toit, a veterinary scientist and Director of Veterinary Services; L.T. Nel representing the Geological Survey; A.R. Saunders from the Department of Agriculture; John Smeath-Thomas, who specialized in inorganic chemistry and was Master of Rhodes University, S.F. Oosthuizen, a radiologist attached to the University of Pretoria who was a substitute for the ubiquitous H.J. van Eck whose new appointment as Director of Food Supplies and Distribution took priority.[61] A final member of the delegation was E.H. Cluver. Eustace Henry Cluver was a key figure in public or preventative health and had served as the director of medical services for the UDF and subsequently headed the South African Institute for Medical Research (SAIMR). Under his leadership, the SAIMR conducted research into occupational health and respiration, hence his suggestion for the physiological effects of dust on mineworkers. Like Schonland, Cluver was also on the board of the CSIR.[62]

Beyond Borders

In his analysis of British science within the context of empire, Roy Macleod identified five phases. The first—which may have begun with the very early voyages of exploration was the identification, classification and ordering of the people, plants and animals in the far-flung regions of the world that would eventually succumb to British rule. The second was an extension of the first and fuelled by the insatiable demand of the Industrial Revolution for new resources and markets. Natural resources were identified and acquired for their utility, and scientific posts were established in India, Australia and South Africa to contribute to growing knowledge of geography, astronomy and botany. Soon, British colonies would emulate

[60] "British Commonwealth Scientific Conference: July, 1946", File Q/T 876/1: British Commonwealth Committee Meetings on Post-war Publication of Scientific War Work, KMG, Group 27, Box 138, DOD.

[61] "S.A. Delegates to Empire Science Talks", *Rand Daily Mail*, 7 June 1946.

[62] J.H.S. Gear, "In Memoriam: Eustace Henry Cluver", *South African Medical Journal*, Vol 62, 24 July 1982, p144.

the metropole in their own establishment of scientific institutions while adopting a certain "deference" to the centrality and authority of British-based institutions. The third phase in the 1870s was marked by a greater sense of independence and incipient nationhood where scientific conferences were held outside the metropole and colonies forged links with each other based on the sharing of knowledge. With both the implementation of high imperialism and the challenges associated with it from the late nineteenth century, science and technology became a less overt means of asserting British dominance and western values. It is in the early twentieth century that "coordination" became the key descriptor of the relationship between Britain and the Dominions and colonies, suggestive of the "family" with Britain the parent to her colonial children and connected by the technologies of communication and transport. With the end of the First World War came the fifth phase: the recognition that the Dominions and the colonies were increasingly equal producers of scientific knowledge yet, even within a growing sense of nationalism, "loyalty" to the empire and the unquestioning assumptions of science, development and progress were retained.[63]

As Edgerton argues, "internationalism" had been a feature of the War, with relationships that were not necessarily based on inequality but marked by the effective sharing of resources between Britain, the Allies and the Dominions. Britain had been intimately connected to the empire through air, sea and communications networks, with the movement of food, supplies and people a defining feature of the War. And it was this support provided by the rest of the world that permitted Britain to continue the struggle against the Axis, as Europe fell. This internationalism was also evident in the training for combat with RAF pilots trained in Canada.[64] The Second World War also demonstrated the value of collaboration, not just between Britain and the Commonwealth but with the United States an increasingly dominant force in the development of science and wartime technology. The success of collaboration was

[63] Refer to Roy Macleod, "Passages in Imperial Science: From Empire to Commonwealth", *Journal of World History*, Vol 4, No 1, 117–150, 1983.

[64] David Edgerton, *Britain's War Machine: Weapons, Resources and Experts in the Second World War* (Penguin Books, 2012), pp5, 24, 51–56. With the implementation of the British Commonwealth Air Training Plan, RAF pilots were trained in Canada, Southern Rhodesia (Zimbabwe) and South Africa. Refer to Derek Wilkins, "RAF Training in Southern Africa", 25 July 2003, *WW2 People's War*, https://www.bbc.co.uk/history/ww2peopleswar/stories/37/a1122337.shtml, accessed 1 May 2025.

manifest in radar and the atomic bomb and it was this sense of accomplishment that influenced the post-war vision of scientific achievement. The "exchange of science of technology" marked the evolving relationship between Britain and the Commonwealth and, as Macleod argues, "imperial science sustained imperial unity [and enabled] imperial diplomacy".[65] Yet, while described in glowing terms as a "People's War", the successful use and allocation of resources that allowed for the successful pursuit of the War was dominated by Britain, the United States and to an extent, Canada.[66] Yet, in the spirit of "internationalism", the Commonwealth Science Conference in 1946, under the auspices of the Royal Society, portrayed "Empire [as an] example to the world of what might be achieved by friendly intellectual discourse".[67]

The list of delegates attending the Official Conference (following on the back of the Royal Society conference) was a lengthy one. The chair and the head of the British Department of Scientific and Industrial Research was Edward Appleton who led the delegation representing Great Britain and Northern Ireland that consisted of figures in trade, agriculture and medicine as well as representatives of colonial offices: J. Gordon represented the India and Burma Office and J.L. Simmons the "Colonial Delegation". The Dominion delegations were also led by the presidents of their national research councils—C.J. Mackenzie for the National Research Council of Canada; David Rivett, his Australian counterpart; E. Marsden for New Zealand; and B.F.J. Schonland for South Africa. After initial consternation as to whether India would send a representative, the director of Scientific and Industrial Research, S.S. Bhatnagar led a delegation representing geology, fisheries and agriculture and also included D.N. Wadia, the president of the National Institute of Sciences. Smaller delegations were received from Burma (with just one representative from the Burma Forest Service), Southern Rhodesia, West Africa, East Africa, Ceylon (a government minerologist), Hong Kong (the Secretary for Development), Palestine and the West Indies. For the larger delegations, figures were from tertiary institutions as well as various

[65] Roy Macleod, "'All for Each and Each for All': Reflections on Anglo-American and Commonwealth Scientific Cooperation, 1940–1945", *Albion: A Quarterly Journal Concerned with British Studies*, Vol 26, No 1, 79–112, 1994, pp80–81.

[66] Edgerton, *Britain's War Machine*, pp150–152.

[67] Quoted in Macleod, "The Royal Society and the Commonwealth", p141.

research centres and organizations.[68] For the colonies, their participation was considered important and the smaller contingents from Ceylon and the West Indies were included due to the "abundance of good will towards the colonies, and [the] desire to help them on the scientific side". Place was then reserved for the colonial administrators and scientists involved in the governance of British territories in Africa, the Middle East, Asia and the Caribbean.[69] However, as the colonies had not reached the levels of scientific development and research as the Dominions (and, to an extent, India), they were seen less as co-creators of knowledge than as the recipients of it.

The conference was opened by Labour politician, Herbert Morrison, who had served as Minister of Supply and subsequently Home Minister under Winston Churchill during the War.[70] In his address, Morrison suggested a "guiding principle", the ultimate aims of the conference and the best means of achieving those. He cautioned against an emphasis on bureaucracy that would consume and neglect the valuable expertise of an already diminished number of "scientific men", and spoke of the importance of distinguishing between research that could be carried out nationally by "the industrial countries of the Empire" and that which required Commonwealth collaboration.[71]

For the Working Committee considering the aims of the conference, Commonwealth collaboration was proposed in terms of the standardization of weights and measures, in electrical and other scientific standards such as that of the wavelength of light and at a basic level, a uniformity in "screw-threads". Other areas of collaboration included the study of radio waves in the atmosphere, already initiated in the pre-war years by New Zealand and Australia; Commonwealth mineral resources and their applicability to British research; and the field of hydrology focusing on

[68] British Commonwealth Scientific Official Conference, London 1946: Report of Proceedings, November 1946, C&O 85/46 - Post-War Scientific Collaboration: Official Civil Science Panel and the Commonwealth Scientific Conference Working Party, IOR/L/E/8/4759, BL.

[69] G.H. Gater, 25 September 1945, C&O 85/46 - Post-War Scientific Collaboration: Official Civil Science Panel and the Commonwealth Scientific Conference Working Party, IOR/L/E/8/4759, BL.

[70] "Herbert Morrison", *Oxford Reference*, https://www.oxfordreference.com/display/10.1093/oi/authority.20110803100210747, accessed 6 June 2024.

[71] British Commonwealth Scientific Official Conference, London 1946: Report of Proceedings, November 1946.

erosion, the movement of water and the silting of waterways. In the last, it was believed that the expertise of the Commonwealth—and in particular, India—was greater than that of Britain due to practical experience. Related to this were concerns regarding erosion, conservation and the proper usage of land across the empire. Collaboration also related to the networks of empire and the distribution of resources, with concerns about the importation of food already infested with insects upon their arrival in British ports and the need to find a solution at the source viz. where the crop was harvested.[72] The partnership between Britain and the Dominions, then reflected specifically British interests as well as an acknowledgement of expertise outside the metropole.

In a context of national self-interest coupled with cross-border co-operation, Morrison's conclusion was a stirring articulation of the optimism of science for an equitable post-war world: "...we are determined that science shall play its proper role in the formation of policy and that its results shall be applied in improving the standard of life, both of the people in this country and of the whole Commonwealth". Scientists were heroically portrayed as "making the world a better and safer place for the people".[73]

In his opening statement, Appleton—who would win the Nobel Prize for Physics a year later—listed the main research areas to make the best use of the limited "scientific manpower". Key was "Teaching and fundamental research" followed by "Civil Science, both government and industrial" and, in a move away from the research priorities that had characterized the preceding six years, "Defence Science" was relegated to third position. He also addressed the role of the university as being the most important site of research, crediting the scientific and technological developments of the War to academics who had joined the war effort, bringing with them the "mental adventurousness and lively imagination so necessary for scientific progress". Yet Appleton also considered the important role of government-sponsored research organizations, with their immense resources, urging each country to find the balance between academia

[72] "Commonwealth Scientific Conference, July 1946: Statements and Memoranda Prepared by the Working Party Appointed by the Civil Science Panel", January 1946, C&O 85/46 - Post-War Scientific Collaboration: Official Civil Science Panel and the Commonwealth Scientific Conference Working Party, IOR/L/E/8/4759, BL.

[73] British Commonwealth Scientific Official Conference, London 1946: Report of Proceedings, November 1946.

and government-sponsored research that suited them best. And he spoke too of the "Commonwealth Patent" as put forth by the UDF via the South African delegation. However, little would come of it. The notion of a "commonwealth patent" had been considered two decades earlier and, while the Conference noted its advantages, it was deemed to be associated with "technical difficulties", making it "impracticable". It was recommended instead that it be raised by the various delegates with their respective governments.[74]

A significant point of discussion on the conference agenda was "Regional Research in Africa" and, fittingly, Schonland was nominated as the chair of the committee that would recommend the importance of considering African challenges on a "regional" rather than "territorial" basis.[75] At the conference, Schonland spoke of the importance of international collaboration in science. A prospective project to which he alluded was a Commonwealth expedition to carry out research in the Antarctic. It was a project initially proposed by the South African Geological Society but Prime Minister Smuts had baulked at South Africa bearing sole financial responsibility for it and was only willing to countenance the project should it be a joint effort.[76] While Schonland spoke of the possible meteorological value of the expedition, it had also attracted the attention of South African geologist Alex du Toit who saw its potential as bolstering the evidence for his then revolutionary advocacy for continental drift theory.[77] For Schonland, science had a global role to play; it exceeded the nationalism of "political boundaries". His address related specifically to the African continent and the need for an organization to co-ordinate scientific research and make recommendations that would promote the development of sub-Saharan Africa. The acquisition of knowledge and the fostering of development was actually limited by territorial borders and Scholand imagined a system of collaboration that was based on the expertise of local—and especially South African—scientists as well as their British counterparts with the improvement of the

[74] British Commonwealth Scientific Official Conference, London 1946: Report of Proceedings, November 1946.

[75] British Commonwealth Scientific Official Conference, London 1946: Report of Proceedings, November 1946.

[76] "S.A. May Take Part in Antarctic Expedition", *Rand Daily Mail*, 29 June 1946.

[77] Refer to Suryakanthie Chetty, *Africa forms the Key: Alex du Toit and the History of Continental Drift* (Cham, Palgrave Macmillan, 2021).

continent being as much a concern of the Colonial Office as it was for those living in Africa.[78] Schonland's vision was realized with the African Regional Scientific Conference held three years later.

"The Beginning of a Great Experiment:"[79] The African Regional Scientific Conference (1949)

The national election of 1948 led to the defeat of Jan Smuts and ushered in a new era in South African politics: the dominance of Afrikaner nationalism and the implementation of apartheid that would lead to South Africa's increasing international isolation and its withdrawal from the Commonwealth in 1961. In 1949 however, the state was more circumspect; as Dubow discusses, during the early years of the National Party government, South Africa continued its involvement in international scientific organizations, both internationally and regionally. Scientific organizations and conferences became the means by which South Africa could maintain international relations.[80] Yet, in his opening address of the African Regional Scientific Conference the new Union Prime Minister D.F. Malan, claimed that the assembly of scientists was free from ideology, politics and state censure, and their aim would be the finding of solutions to uniquely African challenges.[81]

The conference—an outcome of the Commonwealth Scientific Conference three years earlier—was organized by the Deputy President of the CSIR, P.J. du Toit, and was held at the University of the Witwatersrand between 17 and 28 October 1949. The most ambitious scientific conference to be held in South Africa, it was attended by seventy-five delegates representing the colonial states who held territories in Africa as well as thirty-six South African delegates.[82] Papers were submitted beforehand, to be printed and distributed to conference attendees and the first two days were scheduled for presentations by delegates regarding

[78] "Schonland Wants Research Body for Africa", *Rand Daily Mail*, 29 June 1946.

[79] "'If Science Fails Humanity May Have to Evacuate Africa'", *Rand Daily Mail*, 19 October 1949.

[80] Saul Dubow, *A Commonwealth of Knowledge* (Oxford, Oxford University Press, 2006), p257.

[81] "Malan Opens African Regional Scientific Conference", *Rand Daily Mail*, 18 October 1949.

[82] Kingwill, *The CSIR*, p62.

the state of scientific research in their particular regions. Following this, the conference was divided into six areas of discussion ranging from the environment and conservation to medicine and technology. The final two days of the conference would be a summing up of the recommendations to be forwarded to the relevant governments.[83]

Basil Schonland, already fielding prestigious job offers from Britain, found himself in the unenviable position of having to offer an arm to the ailing Malan as they mounted the steps to the Great Hall at Wits where the opening was to take place. With his internationalist leanings, Schonland represented the utopian vision of science[84]—its role in development and the creation of the "world mind". Malan would lead South Africa down a path eventually leading to the privileging of science and technological autarchy in the furtherance of repressive apartheid policy. But that lay in the future. To the unanimous acclaim of attendees, the South African delegation proposed a plan (drawn up by Schonland) for the creation of a scientific council for sub-Saharan Africa. The perceived role of the council articulated the aims of the Commonwealth Scientific Conference (but with the focus on Africa): the co-ordination of research and specialists across the continent, the availability of training, libraries and technical equipment, the co-ordination of the work of the various scientific organizations (such as the CSIR) and the forwarding of "recommendations to the metropolitan governments" for their consideration and implementation. Under the proposed chairmanship of P.J. du Toit (who would replace Schonland as head of the CSIR), it was an illustration of South African aspirations to leadership in science, technology and development in sub-Saharan Africa.[85]

The first meeting of the newly established Scientific Council for Africa South of the Sahara (CSA) was held in Kenya in late 1950. It would ultimately comprise fourteen members and approximately twenty alternate members, all specialists in their fields. These members were drawn from the metropolitan colonial states—France, Portugal, Belgium and Britain—and the white settler states such as South Africa and Rhodesia.

[83] "African Regional Scientific Conference", *South African Journal of Science,* October 1949, p92.

[84] Austin, *Schonland*, pp386–387.

[85] "Scientists Ask with One Voice for Africa Council of Big Scope", *Rand Daily Mail*, 28 October 1949.

Meetings were held annually in various African capitals, addressing scientific issues and with special focus on developmental issues such as environment, conservation, education and nutrition. Smaller meetings were held more frequently with more specific challenges of a technical nature such as infrastructure and water treatment. Fifteen years later however, and in a climate of decolonization, the CSA was perceived to be "[tainted by] colonialism" by independent African states and "rejected".[86] Like the Commonwealth Scientific Conference and the African Regional Scientific Conference, its perceived universalism had been nonetheless predicated on the exclusions and inequalities evident within the context of empire.

In April 1945, Schonland met with Daniel Francois Malan, the leader of the National Party. An Afrikaner nationalist who would become the first Prime Minister of the apartheid state in 1948, Malan believed that the CSIR should have "'n Suid-Afrikaanse kleur".[87] With the rise to power of the National Party, the Minister of Defence sought to reduce UDF ties to Britain[88] and South Africa would eventually leave the Commonwealth in 1961. By then, Schonland had already left South Africa in 1954, to take up a position at the Atomic Energy Research Establishment dedicated to nuclear research in Harwell, England.[89] His new role was symbolic of the world order that had come into being soon after the end of the Second World War. Just as South Africa became increasingly insular and would harness science and technology to national security, the world itself had become polarized into two blocs where H.G. Wells's "world mind" was once again fragmented by power, ideology and war.

At the end of the Second World War, novelist George Orwell was in Germany. Viewing the devastation wrought by aerial bombing raids, Orwell found himself in "actual doubt about the continuity of civilization".[90] Published four years later, *1984* was contextualized by the Second World War and the Cold War that followed soon after. In the novel Orwell evoked a dystopian world, a perversion of the "world mind".

[86] Kingwill, *The CSIR*, pp63–66.

[87] Translated as "a South African colour".

[88] Brian Austin, "The South African Corps of Scientists", *Military History Journal*, Vol 14, No 1, 2007, http://samilitaryhistory.org/vol141ba.html, accessed 2 May 2025.

[89] Austin, *Schonland*, p449.

[90] George Orwell, "Future of a Ruined Germany (1945)", *Essays* (London, William Collins, 2023), p100.

It describes a future where the world is divided into blocs, each of which is at war with the other in a conflict that is unceasing. The result is the creation of Big Brother, the ultimate exertion of power not just on the bodies but on the minds of people, preventing individuality and freedom of thought. All intellectualism gives way to power where there is "no art, no literature, no science", and there is a rejection "of the stupid hedonistic Utopias". Orwell's nightmarish world is based on the manipulation of knowledge—the reduction of language, isolation from the rest of the world, the manipulation of history and of memory, the rejection of scientific discovery.[91] *1984* is a bleak vision where perpetual war creates a static society marked by fear and obedience, the end of progress and of humanity. It is a representation of the new world order that Orwell feared was coming into being, the end point of a century of war.

[91] George Orwell, *Nineteen Eighty-Four* (London, William Collins, 2021), pp226, 227, 281–282, 283–284.

Source List

Archival, Newspaper Sources and Periodicals

Brenthurst Library

Brenthurst Papers, Jack Penn, ed, Johannesburg, Witwatersrand University Press, 1944, Brn. No: 3297, Call No: 617.95 BRE.

Paes, A., "Presentation of a New Stretcher for Cutting Skin Grafts (Skin-stretcher Nobre Leitao)", *Brenthurst Papers,* Jack Penn, ed, Johannesburg, Witwatersrand University Press, 1944, Brn. No: 3297, Call No: 617.95 BRE.

Penn, J., "A Surgeon's Story: It is the Divine Right of Man to Look Human" (unpublished manuscript), Brn No: 12057, 617.95 PEN.

Penn, J. and Brown, L., "A New Conception in Extra-Oral Splintage", *Brenthurst Papers*, Jack Penn, ed, Johannesburg, Witwatersrand University Press, 1944, Brn. No: 3297, Call No: 617.95 BRE.

Major Penn, J., Major Brown, L.J., Captain Berry, T.B., Sgt. Schulmeister, A., Sgt. Roylance J. and C.L. Ormond, C.L., "Problems in Eye Lid and Socket Reconstruction", *Brenthurst Papers,* Jack Penn, ed, Johannesburg, Witwatersrand University Press, 1944, Brn. No: 3297, Call No: 617.95 BRE.

Department of Defence Archives

File 104/34, Medical Boards, Chief Staff Officer (CSO), Box 63.

File 129/376: Plastic Surgery: MOs Studying in England, Director General Medical Services, Box 92.

S. Chetty, *Science and Technology in South Africa, 1939-1946*, Palgrave Studies in the History of Science and Technology,
https://doi.org/10.1007/978-3-032-05084-7

File 320: Minutes of Weekly Meetings of Inventions Sub-Committee, KMG, Group 27, Box 178.
File 320/1: Minutes of Meeting etc Main Inventions Committee – Mr W.E. John's Personal Copies, KMG, Group 27, Box 178.
File 332/1: 1) Brenthurst Clamp Splint 2) The Acrylic Artificial Eye – Brenthurst Inventions, KMG, Group 27, Box 163.
File 334/6: Smooth Bore Gun – Inventor: "Teivos" (Soviet), KMG, Group 27, Box 163.
File 334/9: Detachable Curvature Barrel Extension for Guns – Inventor: W.A. Gaunt, KMG, Group 27, Box 163.
File 380: Body Armour to Protect Soldiers and Airmen (Bullet Proof Clothing) – General Suggestions, KMG, Group 27, Box 165.
File 393: Training of Natives on Military Lines – Inventor: W.H. Blower, KMG, Group 27, Box 178.
File 393/1: Training of Natives for Defence of South Africa – Inventor: S.N. Buiers, KMG, Group 27, Box 178.
File 394: Military Training for Women – Suggester: Kathleen Christensen, KMG, Group 27, Box 178.
File 416: Training of Air Gunners: Synthetic Trainers – Inventor: Air/Sgt L. Melman, KMG, Group 27, Box 167.
File 462: Jet Propulsion – Inventor: R.W. Bullock, KMG, Group 27, Box 169.
File A-1: R.K. Allan - Rocket Buoys, KMG, Group 27, Box 179.
File A3: Inventions Outstanding – Under Investigation as at 20–8–1940, KMG, Group 27, Box 178.
File A.G. 211/4/2: Witwatersrand Agricultural Shows, Adjutant General 3, Box 66.
File B4: Mr Robert E. Booth – Artificial Sandstorms, KMG, Group 27, Box 178.
File B.I. 29: "War Train National Tour", Box 277.
File C6: J. Cronje – Aalwyn and Magawa Trees as Food for Soldiers, KMG, Group 27, Box 178.
File C.G.S. 391/56/1, Vol 4: "Co-operation with Ground Troops and Navy; Co-operation with A.C.F. Units", Group 2, Box 260.
File DTS 19/238: Inventions – General Anti-Tank Invention, Invention by Mr Weiss, KMG, Group 27, Box 143.
File DTS 405, Vol 2: W.E John Esq, Consulting Engineer and Patent Agent, KMG, Group 27, Box 166.
File DTS/INV 338: Bren Gun – Proposed Improvements to – Deflector for Empty Cases: Cpl J.A. Killoran, KMG, Group 27, Box 163.
File DTS/INV 413: Arms for WAAS – Proposer: A Fawcus, Dunstanburgh, Eston, Natal, KMG, Group 27, Box 166.
File H-1: Mr Albert Haskin, KMG, Group 27, Box 179.
File MD 4/4/181, Director Medical Services Group 1, Box 45

File MD 4/5/223, Director Medical Services Group 1, Box 47.
File MD 14/1 (Vol 1): Military Hospital Voortrekkerhoogte General, Surgeon General, Group 2, Box 111.
File MD 14/1/2 (Vol 2): Military Hospital Voortrekkerhoogte Establishment, Surgeon General, Group 2, Box 110.
File MD 14/1/2 (Vol 2): "Military Hospital Voortrekkerhoogte Establishment", Surgeon General, Group 2, Box 115.
File MD 14/1/9: "Military Hospital V.T.H. arrangements re admission of patients for operative and medical treatment", Surgeon General, Group 2, Box 109.
File MD 14/1/14: Military Hospital – Voortrekkerhoogte: Consulting Physicians, Surgeon General, Group 2, Box 108.
File MD 540/6 MVS 615/42: "Brenthurst Splints", GG, Group 2, Box 1101.
File MD 580: Treatment of Facial Injuries, Surgeon General, Group 2, Box 1135.
File MD 580/2: Tara: Red Cross Hospital for Plastic Surgery – General, Surgeon General, Group 2, Box 1136.
File MD 580 (Vol 2): Treatment of Facial Injuries Maxillo-Facial Treatment, R.A.M.C, Group 2, Box 1101.
File MD 580 (Vol 2): Treatment of Facial Injuries Maxillo Facial Treatment, Surgeon General, Group 2, Box 111.
File Q(T) 19/234, Vol 1: Inventions – Destruction of Mine Fields by Major A.S.J. Du Toit, KMG, Group 27, Box 143.
File Q/T 876/1: British Commonwealth Committee Meetings on Post-war Publication of Scientific War Work, KMG, Group 27, Box 138.
"Biesheuvel, Simon", Service Card No W.R.12/7542, 203333V.
"Brown, Leslie Julius", Service Card No W.R.12/7091, 176921.
Gaza, Elias Service File, Regt No 13948.
"Girling, Leslie Alfred", Service Card No W.R.12/2064, 123668V.
"Lempke, Maxwell William", Service Card, 227151.
"Melman, Leonard", Service Card, 208793V.
Penn, Jack Service File, W.R. 12/6640.
Petersen, Norman Personnel File Reg No 16181.
Robertson, Lewis Stevens Service File, W.R. 12/1284, DOD.
"Record of Service of George Clement Gie Werdmuller", File No R/6700: Werdmuller G.C., 87352V.
The Nongqai
– Behrens, H.P.H., "Seventy-Five Air School: the Kindergarten of the Air Force in South Africa", July 1943.
– Douglass, P., "Progress of Modern Warfare: The Army in Review", March 1944.
– Douglass, P., "Cape Town Cavalcade in Retrospect", May 1944.

– Macfarlane, E., "Psychology in the Air Force", July 1943.
– Rowe, J.O., "Psychological Tests for Air Crews", July 1942.
– Sack, J., "The Background to Air Power", March 1944.
– "Something Like Hitler's Secret Weapon!", April 1940.
– "The Union at War: How to Join the Forces", June 1940.

National Archives of South Africa

Ref DWS1045, Vol 146: "Defence Exhibit Rand Show", DGS.

University of the Witwatersrand Central Records Office

Simon Biesheuvel, "Air Personnel Research in South Africa", Box I4: Notes on Various Topics in Psychology, Professor Simon Biesheuvel Collection, 1933–1991, WaB-02-Bie-S.

The National Archives (Kew)

"Committees (Code 18(C)): War Office Committee of Enquiry into Causation and Prevention of 'Shell-Shock'; minutes and correspondence report", 1922, WO 32/4748.

"War Cabinet Committee on Air Organisation and Home Defence Against Air Raids (2nd Report)", 17 August 1917, CAB 24/22/58.

"Commonwealth Scientific Official Conference, July 1946: statements and memoranda prepared by the Working Party appointed by the Civil Science Panel", January 1946, CO 1073/321.

The British Library

British Commonwealth Scientific Official Conference, London 1946: Report of Proceedings, November 1946, C&O 85/46 - Post-War Scientific Collaboration: Official Civil Science Panel and the Commonwealth Scientific Conference Working Party, IOR/L/E/8/4759.

"Commonwealth Scientific Conference, July 1946: Statements and Memoranda Prepared by the Working Party Appointed by the Civil Science Panel", January 1946, C&O 85/46 - Post-War Scientific Collaboration: Official Civil Science Panel and the Commonwealth Scientific Conference Working Party, IOR/L/E/8/4759.

Gater, G.H., 25 September 1945, C&O 85/46 - Post-War Scientific Collaboration: Official Civil Science Panel and the Commonwealth Scientific Conference Working Party, IOR/L/E/8/4759.

Secretaries, Royal Society, 22 January 1945, C&O 85/46 - Post-War Scientific Collaboration: Official Civil Science Panel and the Commonwealth Scientific Conference Working Party, IOR/L/E/8/4759.

United Nations Library and Archives

"League of Nations: international Cooperation", https://libraryresources.unog.ch/lonintellectualcooperation/ICIC, accessed 5 May 2025.
"Annex E: Report on the Organisation of Intellectual Labour" presented by M Lafontaine, 18 December 1920, https://libraryresources.unog.ch/lonintellectualcooperation/ICIC, accessed 5 May 2025.
League of Nations, "International Organisation of Intellectual Work", 21 September 1921, https://libraryresources.unog.ch/lonintellectualcooperation/ICIC, accessed 5 May 2025.

Miscellaneous Archive/Primary Sources

A.S.J. Du Toit, "Apparatus for Clearing Land Mines or Minefields", 2,489,564, United States Patent Office, 29 November 1949.
A.S.J. Du Toit, "Improved Apparatus for Clearing Land Mines or Minefields", Patent Specification 573,753, 24 March 1942 (London, Patent Office, 1946), https://worldwide.espacenet.com/patent/search/family/009767438/publication/GB573753A?q=pn%3DGB573753A, accessed 15 December 2023.
Amateur Mechanics, Chicago, Popular Mechanics, 1907.
"The Atlantic Charter: Declaration of Principles issued by the President of the United States and the Prime Minister of the United Kingdom", 14 August 1941, *North Atlantic Treaty Organization*, https://www.nato.int/cps/en/natohq/official_texts_16912.htm, accessed 23 April 2024.
Letter by G.W.D. Breadon, "Travelling Caterpillar Forts", 19 September 1914, PRG 1319/1/1, State Library South Australia Digital Collections, https://digital.collections.slsa.sa.gov.au/nodes/view/2460, accessed 18 December 2023.
LH Robertson Collection, 7/2/36, Liddell Hart Centre for Military Archives, King's College Archives, London.
Republic of South Africa Government Gazette, No 57 of 1978: Patents Act, 1978, Cape Town, 17 May 1978.
US Military Intelligence Service, War Department, "German Coastal Defences", Special Series No 15, 15 June 1943.

Rand Daily Mail

- 11 October 1940.
- "300 m.p.h Fighter Planes for Union", 3 February 1939.
- Advertisement, 26 July 1947.
- "Air Display in Pretoria on Saturday", 15 October 1941.
- "Air Commando to Stage New Flag Ceremony", 5 August 1941.
- "'Aviation Medicine' Urged", 3 February 1939.
- "The Air Commando is Coming", 2 August 1941.
- "'Beautiful Bombs' to be Exhibited at Liberty Cavalcade", 19 May 1942.
- "Bloemfontein Welcomes Convoy", 7 November 1940.
- "Cape Towns [sic] Welcome Steel Commando", 23 October 1940.
- "Caproni First Plane on New Aerodrome", 16 August 1941.
- "Cavalcade Closes", 2 June 1942.
- "Churchill Sends Message to Scientists", 27 September 1941.
- "City Dentists Invent Revolutionary Surgical Appliance", 1 April 1943.
- "City Needs Home for War Museum", 28 March 1947.
- "City Praised for Stirring Welcome", 11 December 1940.
- "City's Big Welcome to Air Commando",7 August 1941.
- "City's Remarkable Response to Liberty Cavalcade", 2 June 1942.
- "Commando Not a Waste of Money", 23 September 1941.
- "Death of Sir Edward Thornton; Long Service in Public Health", 28 October 1946.
- "Ex-Ossewa Chairman's Letter", 10 September 1941.
- "Gates of Liberty Cavalcade to Open this Morning", 23 May 1942.
- "General Smuts Hopes Old Friends will Return to the Fold", 11 November 1940.
- "Great Army Column to Move Tomorrow", 31 October 1940.
- "Greet the Commando To-day", 10 December 1940.
- "Herenigde Party will Become S.A. Curse", 11 November 1940.
- "His Excellency Sees Great Arms Drive", 18 October 1940.
- "'If Science Fails Humanity May Have to Evacuate Africa'", 19 October 1949.
- "Indians to Give Rare Golden Dagger to Smuts", 22 May 1942.
- "Industrial Board Meets", 18 October 1940.
- "Malan Opens African Regional Scientific Conference", 18 October 1949.
- "Many Inventions", 12 October 1940.
- "Museum Piece", 17 December 1947.
- "New Bureau will Provide Industry Aptitude Tests", 14 January 1946.
- "Nylstroom Council to Shun War Train", 3 April 1941.
- "Premier Praises City for Stirring Welcome to Steel Commando", 11 December 1940.
- "Real Bombs to be Used in C.P.S. Display", 6 August 1941.
- "Recruits' Scroll", 2 November 1940.

– "S.A. Delegates to Empire Science Talks", 7 June 1946.
– "S.A. May Take Part in Antarctic Expedition", 29 June 1946.
– "S.A. Scientists Wanted for Big Conference: British Association to Discuss African Problems", 9 September 1941.
– "Scientists Ask with One Voice for Africa Council of Big Scope", 28 October 1949.
– "Schonland Wants Research Body for Africa",29 June 1946.
– "Schonland Will Share in Empire Atomic Bomb Talks", 2 March 1946.
– "Science Rightly Used Can Abolish War", 3 July 1940.
– "Smuts Believes War Will End after 1944: Africa, Continent of the Future, Destined to Play Big Part", 28 August 1943.
– "Smuts Calls for Great Industrial Research Move", 9 October 1945.
– "S. Africans have High Physical Potential", 15 July 1941.
– "South African Invention Used on Normandy Battlefield", 30 June 1944.
– "Stature May Halt, Brain Will Increase: Dr Broom Visualises Superman", 21 May 1938.
– "Stellenbosch Mayor Joins Up as Private", 25 April 1941.
– "Union's Growing 'Arms Factory' to be on View at Rand Easter Show", 8 March 1940.
– "Vanity Fair", 1 May 1941.
– "Vanity Fair", 5 July 1941.
– "War Museum is Memorial to S.A.'s Greatest Efforts – Smuts", 30 August 1947.
– "War Museum Opened in Zoo Grounds", 10 April 1947.
– "We're not reading history…we're making it!", 10 December 1940.
– "Women Recruits Needed", 8 October 1941.
– "Wreath Laid on Grave of Trek Leader", 10 December 1940.

South African Railways and Harbours Magazine

– "Myers, C.S., "Applied Psychology in War", *Manchester Guardian* reprinted in *South African Railways and Harbours Magazine*, Vol 35, July 1941.
– Vol 34, April 1940
– Vol 35, August 1941
– Vol 35, September 1941
– Vol 36, February 1942
– Vol 36, April 1942
– Vol 36, May 1942

Secondary Sources: Books, Book Chapters, Journal Articles and Online Sources

"£20,000 Award for Inventors of Flail Tanks", *The Times*, 2 September 1948, https://trove.nla.gov.au/newspaper/article/78590436, accessed 15 December 2023.

"The Activities of the Society", *Notes and Records of the Royal Society of London*, Vol 4, No 1, April 1946.

Adas, M., *Machines as the Measure of Men: Science, Technology and Ideologies of Western Dominance*, Ithaca and London, Cornell University Press, 2014.

"African Regional Scientific Conference", *South African Journal of Science*, October 1949.

Ahrenfeldt, R.H., *Psychiatry in the British Army in the Second World War*, London and New York, Routledge, 1958.

Alexander, C., "Faces of War" in *Smithsonian Magazine*, February 2007, https://www.smithsonianmag.com/arts-culture/faces-of-war-145799854/, accessed 12 August 2020.

Alfonso, K.L.M., *Femme Fatale: An Examination of the Role of Women in Combat and the Policy Implications for Future American Military Operations*, Air University Press, 2009.

"The Army Collection of War Specimens: The Opening of the Exhibition", *The Lancet*, 20 October 1917.

Asuelime, L., "Uranium politics of gatekeeping: Revisiting the British Government's Policy vis-à-vis South Africa, 1945–1951", *Historia*, Vol 58, No 1, 33–50, 2013.

Austin, B., *Schonland: Scientist and Soldier*, Bristol and Philadelphia, Institute of Physics Publishing, 2001.

Austin, B., "The South African Corps of Scientists", *Military History Journal*, Vol 14, No 1, 2007, http://samilitaryhistory.org/vol141ba.html, accessed 2 May 2025.

"Australia to the Fore: Invention of the War Tank", *The Register*, 12 February 1920, https://trove.nla.gov.au/newspaper/article/65020239, accessed 18 December 2023.

"The Awards Made to Tank Designers by the Royal Commission", *The Spectator*, 6 December 1919, http://archive.spectator.co.uk/article/6th-december-1919/3/the-awards-made-to-tank-designers-by-the-royal-com, accessed 18 December 2023.

Aymard, J.L., "Nasal Reconstruction: The Tubed Pedicle Flap", *The Lancet*, 27 April 1935.

Aymard, J.L., "Nasal Reconstruction with a note on Nature's Plastic Surgery", *The Lancet*, 15 December 1917.

Bailey, M.S., "A brief history of British military experiences with infectious and tropical diseases", *J R Army Med Corps*, 159, 150–157, 2013.

Ballandonne, M., "The History of Futures Studies: A Note on Gilfillan's Early Work", *Technological Forecasting and Social Change*, Vol 157, August 2020.

Bamji, A., *Faces from the Front: Harold Gillies, The Queen's Hospital, Sidcup and the Origins of Modern Plastic Surgery*, Solihull, Helion and Company Ltd, 2017.

E.H. Beardsley, "Secrets between Friends: Applied Science Exchange between the Western Allies and the Soviet Union during World War II", *Social Studies of Science*, Vol 7, No 4, 447–473, 1977.

Beinart, W., "Soil Erosion, Conservationism and Ideas about Development: A Southern African Exploration, 1900–1960", *Journal of Southern African Studies*, Vol 11, No 1, 52–83, 1984.

Beinart, W. and Dubow, S., *The Scientific Imagination in South Africa: 1700 to the Present*, Cambridge, Cambridge University Press, 2021.

Biesheuvel, S., *African Intelligence*, Johannesburg, South African Institute of Race Relations, 1943.

Bigg, C., Reinisch, J., Somsen, G. and Widmalm, S., "The Art of Gathering: Histories of International Scientific Conferences", *The British Journal for the History of Science*, 56, 2023.

Bird, K. and Sherwin, M.J., *American Prometheus: The Triumph and Tragedy of J. Robert Oppenheimer*, London, Atlantic Books, 2009.

Bourke, J., *An Intimate History of Killing: Face-to-Face Killing in Twentieth-Century Warfare*, London, Granta Books, 1999.

Bourke, J., *The Second World War: A People's History*, New York, Oxford University Press, 2001.

Bradley, J., *Flyboys: A True Story of Courage*, New York, Back Bay Books, 2004.

Broom, B., "The Boskop Skull" (Letter to the Editor) in *Nature*, No, 2929, Vol. 116, 19 December 1925, p897.

Burgess, W., "Foreword", *Scientia Militaria*, Vol 50, No 2, v-xv, 2022.

Butler, S., *East to the Dawn: The Life of Amelia Earhart*, Cambridge, Da Capo Press, 1999.

Campbell-Smith, D., *Jet Man: The Making and Breaking of Frank Whittle, Genius of the Jet Revolution*, London, Head of Zeus Ltd, 2021.

Chetty, S., *Africa forms the Key: Alex du Toit and the History of Continental Drift*, Cham, Palgrave Macmillan, 2021.

Chetty, S., "Our Victory was Our Defeat: Race, Gender and Liberalism in the Union Defence Force, 1939–1945", University of KwaZulu-Natal, Unpublished PhD Dissertation, 2006.

Chetty, S., *Reconstructive Surgery and Modernisation in Twentieth-Century South Africa: The Professional and Public Life of Jack Penn*, Chaim, Palgrave Macmillan, 2023.

Chetty, S., "'Subjects or Citizens?': Black South Africans and the Dilemma of the Second World War", *Journal of Natal and Zulu History*, 53–79, Vol 30, No 1, 2012.

Chetty, S., "'A White Man's War': Settler Masculinity in the Union Defence Force, 1939–1945", *Africa and World War II,* J.A. Byfield, C.A. Brown, T. Parsons and A.A. Sikainga, eds, New York, Cambridge University Press, 2015.

Collins, P.M.D., "The British Association for the Advancement of Science and Public Attitudes to Science, 1919–1945", University of Leeds, unpublished PhD dissertation, 1978.

"Conference on Science and World Order", *Nature*, 4 October 1941, Vol 148.

Cook, C., "A Fascist Memory: Oswald Mosley and the Myth of the Airman", *European Review of History*, Vol 4, No 2, 147–162, 1997.

Cornwell, J., *Hitler's Scientists: Science, War, and the Devil's Pact*, Penguin, 2003.

Daries, A.S., "The History of Physical Education at Stellenbosch University", Stellenbosch University, Unpublished PhD Dissertation, 2023.

Davenport, T.R.H., *South Africa: A Modern History*, Johannesburg, Macmillan South Africa, 1978.

Dawson, R.L.G., "The History, Antecedents and Progress of the Mount Vernon Centre for Plastic Surgery and Jaw Injuries, Northwood, Middlesex 1939–1983", *British Journal of Plastic Surgery*, 41, 83–91, 1988.

Dedering, T., "Air Power in South Africa, 1914–1939", *Journal of Southern African Studies*, Vol 41, No 3, 451–465, 2015.

Dedering, T., "Blue Skies into White Space: Southern African Responses to the Trans-African Flight of the *Silver Queen*, 1920", *Technology and Culture*, Vol 59, 289–312, April 2018.

Dedering, T., "'Prophet of Air Power': Jan Smuts and the History of Aviation", Michael Eckhardt, ed., *Mission Afrika. Geschichtsschreibung über Grenzen hinweg. Festschrift für Ulrich van der Heyden*, Stuttgart, Franz Steiner, 2019.

Dedering, T., "South Africa's Secret Chemical Weapons Project, 1933–1945", *The International History Review*, Vol 44, No 2, 263–281, 2022.

Doherty, R., *Hobart's 79th Armoured Division at War: Invention, Innovation and Inspiration*, Barnsley, Pen and Sword Books Ltd, 2011.

Dubow, S., *A Commonwealth of Knowledge: Science, Sensibility and White South Africa 1820–2000* (New York, Oxford University Press, 2007.

Dubow, S., "Mental Testing and the Understanding of Race in Twentieth-Century South Africa", *Science, Medicine and Cultural Imperialism*, Teresa Meade and Mark Walker, eds, Basingstoke, Macmillan, 1991.

Dumett, R., "Africa's Strategic Materials during the Second World War", *The Journal of African History*, Vol 26, No 4, 381–408, 1985.

Edgerton, D., *Britain's War Machine: Weapons, Resources and Experts in the Second World War*, Penguin Books, 2012.

Edgerton, D., "The Contradictions of Techno-Nationalism and Techno-Globalism: A Historical Perspective", *New Global Studies*, Vol 1, No 1, 2007, 1–32.

Edgerton, D., *England and the Aeroplane: Militarism, Modernity and Machines*, Penguin, 2013.

Edgerton, D., *The Shock of the Old: Technology and Global History Since 1900*, London, Profile Books, 2008).

Edgerton, D., *Warfare State: Britain, 1920–1970*, Cambridge, Cambridge University Press, 2006.

Einstein, A., "Science and Civilisation", *Essays in Humanism*, New York, Open Road, 2015.

Els, P.J., *Chronicles of the Heights, Vol 1*, Valhalla, Pelsa Books, 2012.

Evans, J., "History of the School of Aviation Medicine, 1918–1958", *The Air Power Historian*, Vol 5, No 4, 245–261, October 1958.

Fessenden, M., "The History of Military Parades in the US", *The Smithsonian Magazine*, 7 February 2018, https://www.smithsonianmag.com/smart-news/how-military-parades-us-have-changed-180968102/, accessed 10 June 2024.

"Follow the Drum: South African Military Hospital", *The Library Blog*, 8 November 2017, https://libraryblog.lbrut.org.uk/2017/11/south-african-military-hospital/, accessed 25 November 2020.

Franssen, B.G.M., Schuurman, A.H., Aebele, Van Der Molen, M. and Kon, M., "One Century of Kirschner Wires and Kirschner Wire Insertion Techniques: A Historical Review", *Acta Orthopaedica Belgica*, 76, 1–6, 2010.

French, D., *Military Identities: The Regimental System, the British Army and the British People c 1870–2000*, Oxford, Oxford University Press, 2005.

Freund, B., "The South African Developmental State and the First Attempt to Create a National Health System: Another Look at the Gluckman Commission of 1942–1944", *South African Historical Journal*, Vol 64, No 22, 170–186, 2012.

Freund, B., *Twentieth-Century South Africa: A Developmental History*, Cambridge, Cambridge University Press, 2019.

Friedman, D.M., *The Immortalists: Charles Lindbergh, Dr Alexis Carrel and their Daring Quest to Live Forever*, New York, Harper Perennial, 2008.

"Full History – USS *New Jersey*: The World's Greatest Battleship", https://www.battleshipnewjersey.org/, accessed 5 October 2024.

Gear, J.H.S., "In Memoriam: Eustace Henry Cluver", *South African Medical Journal*, Vol 62, 24 July 1982.

Gilfillan, S.C., "Inventiveness by Nation: A Note on Statistical Treatment", *Geographical* Review, Vol 20, No 2, 1930.

Gillies, H.D., *Plastic Surgery of the Face*, London, Henry Frowde and Hodder and Stoughton, 1920.

Gleeson, I., *The Unknown Force: Black, Indian and Coloured Soldiers Through Two World Wars*, Rivonia, Ashanti Publishing, 1994.

Grant, R.G., *Flight: The Complete History of Aviation*, London, Dorling Kindersley Limited, 2017.

"Sir Richard Gregory, Bart., F.R.S", *Nature*, Vol 153, 1944.

Grunden, W.E., Kawamura, Y., Kolchinsky, E., Maier, H., and Yamazaki, M., "Laying the Foundation for Wartime Research: A Comparative Overview of Science Mobilization in National Socialist Germany, Japan, and the Soviet Union", *Osiris*, 2nd Series, Vol. 20, 79–106, "Politics and Science in Wartime: Comparative International Perspectives on the Kaiser Wilhelm Institute".

Grundlingh, L., "The Military, Race and Resistance: The Conundrums of Recruiting Black South African Men during the Second World War", *Africa and World War II,* J.A. Byfield, C.A. Brown, T. Parsons and A.A. Sikainga, eds, New York, Cambridge University Press, 2015.

Gunga, H., *Nathan Luntz: His Life and Work in the Fields of High Altitude Physiology and Aviation Medicine*, Elsevier Inc, 2009.

Hacker, B.C. and Vining, M., "Military Museums and Social History", *Does War Belong in Museums?: The Representation of Violence in Exhibitions*, Wolfgang Muchitsch (ed.), (Bielefeld, Transcript Verlag, 2013.

Harries-Jenkins, G., "The Development of Professionalism in the Victorian Army", *Armed Forces and Society*, Vol 1, No 4, 1975.

Harris, J.P., *Men, Ideas and Tanks: British Military Thought and Armoured Forces, 1903–1939*, Manchester, Manchester University Press, 1995, 2015.

Harsch, V., "Heinz von Diringshofen: Aeromedical Pioneer", *Aerospace Medicine and Human Performance,* 501–503, Vol 90, No 5, May 2019.

Harsch, V., "Ludolph Brauer: German Aeromedical Pioneer", *Aviation, Space and Environmental Medicine,* Vol 5, No 8, August 2004.

Hartley, M., "Tropical Ulcer", https://dermnetnz.org/topics/tropical-ulcer, accessed 18 March 2024.

"Haughton, Dr Sidney Henry", *S2A3 Biographical Database of Southern African Science*, https://www.s2a3.org.za/bio/Biograph_final.php?serial=1238, accessed 27 April 2024.

Hemingway, E., *Hemingway on War*, London, Vintage, 2014.

Hemming, H., *The Ingenious Mr Pyke: Inventor, Fugitive, Spy*, New York, Public Affairs, 2015.

"Herbert Morrison", *Oxford Reference*, https://www.oxfordreference.com/display/https://doi.org/10.1093/oi/authority.20110803100210747, accessed 6 June 2024.

Hocking, A., *Oppenheimer and Son*, Johannesburg, McGraw-Hill Book Company, 1973.

Holledge, R., "The Absurdities of War: How Heath Robinson Eschewed Jingoism", *The New European*, 18 August 2022, https://www.theneweuropean.co.uk/the-absurdities-of-war/, accessed 2 January 2024.

Holmes, J.D., "Development of Plastic Surgery", *War Surgery, 1914–1918,* Thomas Scotland and

Hyam, R., and Henshaw, P., *The Lion and the Springbok: Britain and South Africa since the Boer War*, Cambridge, Cambridge University Press, 2003.

Hyslop, J., "'Segregation has fallen on evil days': Smuts' South Africa, Global War, and Transnational Politics, 1939–46", *Journal of Global History*, Vol 7, 438–460, 2012.

Jacobsen, A., *Operation Paperclip: The Secret Intelligence Program that Brought Nazi Scientists to America*, New York, Back Bay Books, 2015.

James, L., *Churchill and Empire: Portrait of an Imperialist*, London, Weidenfeld and Nicolson, 2021.

James, L., "'Now inhale the gas': Interactive Readership in Two Victorian Boys' Periodicals, 1855–1870", *Victorian Periodicals Review,* 64–80, Vol 42, No 1, 2009.

"John Gillespie Magee Jr", *Poetry Foundation*, https://www.poetryfoundation.org/poets/john-gillespie-magee-jr, accessed 6 May 2025.

Jokl, E., *Aviation Medicine*, Cape Town, Unie-Volkspers Beperk, 1942.

Jokl, E., *South African Reminiscences*, Ernst Jokl, 1988.

Josephson, P.R., *Totalitarian Science and Technology*, New York, Humanity Books, 2005.

Kennedy, P., *Engineers of Victory: The Problem Solvers Who Turned the Tide in the Second World War*, Penguin Books, 2014.

Kiladi, M., "A Historical Perspective on Lionel Penrose: Scientist, Geneticist and Dedicated Opponent of Eugenics", *Annals of Human* Genetics, 1–8, 2024.

Kilner, Thomas Pomfret (1890 - 1964), *Plarr's Lives of the Fellows*,
https://livesonline.rcseng.ac.uk/client/en, accessed 25 March 2024.

D.G. Kingwill, *The CSIR: The first 40 years*, Pretoria, CSIR, 1990.

Kinnear, K., "Father of the SAAF", 26 December 2010, https://web.archive.org/web/20111004201645/http://www.saafmuseum.org.za/features/2010/father-of-the-saaf/, accessed 23 September 2024.

Kotzé, E.M., "More Than Just Pretty Girls in Uniform: A Historical Study of Women's Military Roles during World War II, 1939–1945", Stellenbosch University, Unpublished PhD Dissertation, 2021.

Kuljian, C., *Darwin's Hunch: Science, Race and the Search for Human Origins,* Johannesburg, Jacana Media, 2016.

Landes, D., *The Wealth and Poverty of Nations*, London, Abacus, 1998.

Latilla, M., *Johannesburg: Then and Now*, https://johannesburg1912.com/2015/04/29/history-of-braamfontein-pt-4-rand-show-wits-nunnery-brewery/, accessed 6 February 2024.

Leal, P.E., "Guernica", Museo Reina Sofía, https://www.museoreinasofia.es/en/collection/artwork/guernica, accessed 6 May 2025.

MacFarlane, R., *Mountains of the Mind: A History of a Fascination*, London, Granta Books, 2008.

Macleod, R., "'All for Each and Each for All': Reflections on Anglo-American and Commonwealth Scientific Cooperation, 1940–1945", *Albion: A Quarterly Journal Concerned with British Studies*, Vol 26, No 1, 79–112, 1994.

Macleod, R., "Passages in Imperial Science: From Empire to Commonwealth", *Journal of World History*, Vol 4, No 1, 117–150, 1983.

Macleod, R., "The Royal Society and the Commonwealth: Old Friendships, New Frontiers", *Notes and Records of the Royal Society of London*, Vol 64, Supplement 1: The Royal Society and Science in 20th Century, 137–149, 2010.

Macmillan, M., *Paris 1919: Six Months that Changed the World*, London, John Murray Publishers, 2019.

Mangin, G. and Lloyd, S., "The Special Signal Services (SSS): We scanned the seas and skies in the Second World War", *Military History Journal*, Vol 11, No 2, 1998, http://samilitaryhistory.org/vol112ml.html, accessed 7 June 2024.

Marks, S., *Divided Sisterhood: Race, Class and Gender in the South African Nursing Profession*, Johannesburg, Witwatersrand University Press, 1994.

Marks, S., "Reflections on the National Health Services Commission: A Response to Bill Freund and Anne Digby on the Gluckman Commission", *South African Historical Journal*, Vol 66, No 1, 169–187, 2014.

Martin, G., "SANDF to be out in force at Rand Show", 22 March 2024, https://www.defenceweb.co.za/sa-defence/sa-defence-sa-defence/sandf-to-be-out-in-force-at-rand-show/, accessed 24 March 2024.

Marwick, A., *The Deluge: British Society and the First World War*, Basingstoke, Palgrave Macmillan, 1991.

McCormack, R.L., "Man with a Mission: Oswald Pirow and South African Airways, 1933–1939", *Journal of African History*, 543–557, 20, 1979.

McCrea, W., "Richard van der Riet Woolley, 24 April 1906–24 December 1986", *Biographical Memoirs of Fellows of the Royal Society*, 922–982, Vol 34, December 1988.

McCalister, L., "A.A. Milne Poem", 9 September 2016, *The Tank Museum*, https://tankmuseum.org/article/milne-tank-poem-found, accessed 21 December 2023.

McCullough, D., *The Wright Brothers*, London, Simon and Schuster, 2016.

Meikle, M.C., "The Evolution of Plastic and Maxillofacial Surgery in the Twentieth Century: The Dunedin Connection", *The Surgeon*, Vol 4, No 5, 325–334.

Milne, A.A., "Poem", *The Tank Museum*, https://tankmuseum.org/article/milne-tank-poem-found, accessed 21 December 2023.

Moore, H.P. and Glennan, K., *Virginan-Pilot*, 18 December 1903, reproduced in Chiles T.A. Larson, "The Big Story", *Virginia Living*, https://virginialiving.com/culture/the-big-story/, accessed 11 January 2024.

Moran, L., *The Anatomy of Courage*, London, Constable and Robinson Limited, 1945.

Morgan-Owen, D., "A Revolution in Naval Affairs? Technology, Strategy and British Naval Policy in the 'Fisher Era'", *Journal of Strategic Studies*, 944–965, Vol 38, No 7, 2015.

Mouton, F.A., *The Opportunist: The Political Life of Oswald Pirow, 1915–1959*, Hatfield, Protea Book House, 2020.

Mulley, C., *The Women Who Flew for Hitler: The True Story of Hitler's Valkyries*, London, Macmillan, 2017.

Murray, N., *The Rocky Road to the Great War: The Evolution of Trench Warfare to 1914*, Washington DC, Potomac Books, 2013.

Nattrass, N., "Economic Growth and Transformation in the 1940s", Saul Dubow and Alan Jeeves, eds., *South Africa's 1940s: Worlds of Possibilities*, Cape Town, Double Storey Books, 2005.

Ngubane, H., *Body and Mind in Zulu Medicine: An Ethnography of Health and Disease in Nyuswa-Zulu Thought and Practice*, London, Academic Press, 1977.

Nicholl, C., *Leonardo da Vinci: The Flights of the Mind*, London, Penguin Books, 2007.

"Norden M-9 Bombsight", National Museum of the United States Air Force, https://www.nationalmuseum.af.mil/Visit/Museum-Exhibits/Fact-Sheets/Display/Article/196340/norden-m-9-bombsight/, accessed 1 March 2024.

"Norden Bombsight", *Air and Space Magazine*, April 2020, https://www.smithsonianmag.com/air-space-magazine/norden-bombsight-180974472/, accessed 1 March 2024.

"Obituary Notices: T Pomfret Kilner", *British Medical Journal*, 2, 127, 11 July 1964, https://www.bmj.com/content/2/5401/127, accessed 10 December 2020.

Orwell, G., "Future of a Ruined Germany (1945)", *Essays*, London, William Collins, 2023.

Orwell, G., *Nineteen Eighty-Four*, London, William Collins, 2021.

"Our Favourite Stories from 120 Years of 'Popular Mechanics'", 11 January 2022, https://www.popularmechanics.com/about/a38726786/our-favorite-stories-from-120-years-of-popular-mechanics/, accessed 2 March 2024.

Overy, R., *1939: Countdown to War*, Penguin, 2010.

Overy, R., *The Bombing War: Europe 1939–1945*, Penguin Books, 2014.

Paris, M., "The Rise of the Airmen: The Origins of Air Force Elitism, c. 1890–1918", *Journal of Contemporary History*, Vol 28, No 1, 123–141, 1993.

Pawle, G., *The Secret War, 1939–1945*, London, George G. Harrap and Co. Ltd, 1956.

Penn, J., "Impressions of Plastic and Maxillo-Facial Surgery in the Palestine War", *Plastic and Reconstructive Surgery*, 6, 4, 327–333, October 1950.

Penn, K., *Reflections on Life*, Johannesburg, Ernest Stanton Publishers, 1980.

Penn, J., *The Right to Look Human: An Autobiography*, Johannesburg, Hugh Keartland Publishers, 1976.

Percy, W., "Jim Crow, Uncle Sam, and the Formation of the Tuskegee Flying Units", *Social Education*, Vol 63, No 1, 14–21, 1999.

"Petersen, Norman August Marais", *Plarr's Lives of the Fellows*, Royal College of Surgeons of England, https://livesonline.rcseng.ac.uk/client/en_GB/lives/search/detailnonmodal?qu=LIVES_OCCUPATION%3D%22Maxillofacial+surgeon%22&qf=LIVES_HONOURS%09Titles%2FQualifications%09MRCS+1920%09MRCS+1920&d=ent%3A%2F%2FSD_ASSET%2F0%2FSD_ASSET%3A378202~~0&ic=true, accessed 6 May 2021.

Peyton, G., *Fifty Years of Aerospace Medicine: Its Evolution since the Founding of the United States Air Force School of Aerospace Medicine in January 1918*, AFSC Historical Publications Series No. 67–180, 1968.

Phillipson, D.J.C., "The National Research Council of Canada: Its Historiography, its Chronology, its Bibliography", *Scientia Canadensis*, Vol 15, No 2(41), 177–193, 1991.

Philpott, W., *Three Armies on the Somme: The First Battle of the Twentieth Century*, New York, Alfred A. Knopf, 2010.

Pine, K., Brian Sloan, B. and Jacobs, R., "A Brief History of Prosthetic Eyes", https://eyemuseum.ranzco.edu/wp-content/uploads/2019/06/History_V4_2.pdf, accessed 20 March 2024.

Plug, C., "Van der Bijl, Dr Hendrik Johannes", *S2A3 Biographical Database of Southern African Science*, https://www.s2a3.org.za/bio/Biograph_final.php?serial=2909, accessed 6 December 2024.

Pole-Evans, I.B., "Our Aloes: Their History, Distribution and Cultivation", *The Journal of the Botanical Society of South Africa*, 1919.

Pomeranz, K., *The Great Divergence: China, Europe, and the Making of the Modern World Economy*, Princeton University Press, 2001.

Porter, R., *The Greatest Benefit to Mankind: A Medical History of Humanity from Antiquity to the Present*, London, Fontana Press, 1999.

Prakash, R., Gopikrishna, V. and Kandaswamy, D., "Gutta-Percha: An Untold Story", *Endodontology*, Vol 17, No 2, 32–36, 2005.

Prasar, V., "Shanti Swarup Bhatnagar", http://www.vigyanprasar.gov.in/scientists/ssbhatnagar/ShantiSwarupBhatnagar.htm, accessed 24 April 2024.

"Price, Mr Bernard", *S2A3 Biographical Database of Southern African Science*, https://www.s2a3.org.za/bio/Biograph_final.php?serial=2247, accessed 27 April 2024.

Rainsford Mowlem, J.L. Buxton, D., MacGregor, A.B., Barron, J.N., "External Pin Fixation for Fractures of the Mandible", *The Lancet*, 4 October 1941.

Raven, J.C., "Standardization of Progressive Matrices, 1938", *British Journal of Medical Psychology*, 19, 137–150, 1941.

Raven, J. and Raven, J., "John Carlyle Raven, 1902–1970, and His Legacy", http://eyeonsociety.co.uk/resources/J-C-Raven-Biography.pdf, accessed 5 October 2024.

Remarque, E.M., *All Quiet on the Western Front*, Brian Murdoch (translator), Alfred A. Knopf, 2018.

Robinson, D.H., *The Dangerous Sky: A History of Aviation Medicine*, Henley-on-Thams, GT Foulis and Co Ltd, 1973.

Rogers, D., *Top Secret: British Boffins in World War One*, Solihull, Helion and Company, 2013.

Roudebush, M., "A Battle of Nerves: Hysteria and its Treatments in France during World War I", Mark S. Micale and Paul Lerner, eds, *Traumatic Pasts: History, Psychiatry and Trauma in the Modern Age, 1870–1930*, Cambridge University Press, 2001.

"Royal Commission on Awards to Inventors", *Nature*, No. 4013, 28 September 1946.

"Science and the World Mind", *Nature*, 11 October 1941, Vol 148.

Shachtman, T., *Laboratory Warriors: How Allied Science and Technology Tipped the Balance in World War II*, New York, Perennial, 2003.

Singh, M., Barrett Noone, R., Afolabi, H. and Caterson, E.J., "A Tribute to the Founding Figures of the American Board of Plastic Surgery", *Plastic and Reconstructive Surgery*, 135, 6, 1047–1054, June 2015.

Smith, R.W., "The Story of Two 'Joiners': The capture, trial, sentencing and execution of Pieter Bouwer and Adolf van Emmenes", *Military History Journal*, Vol 15, No 5, 2012, http://samilitaryhistory.org/vol155rs.html, accessed 18 November 2024.

Sobel, D., *The Elements of Marie Curie: How the Glow of Radium Lit a Path for Women in Science*, London, 4th Estate, 2024.

Stagner, A.C., "Healing the Soldier, Restoring the Nation: Representations of Shell Shock in the USA during and after the First World War", *Journal of Contemporary History*, Vol 49, No 2, 255–274, 2014.

Steven Heys, eds, Solihull, Helion and Company, 2012.

Sullivan, E.P., "The Eye at War: American Eye Prosthetics During the World Wars", 25 September 2017, *Nursing Clio*, https://nursingclio.org/2017/09/25/the-eye-at-war-american-eye-prosthetics-during-the-world-wars/, accessed 20 March 2024.

Tharoor, I., "Military Parades", *Time*, 1 October 2009, https://time.com/archive/6948154/military-parades/, accessed 10 June 2024.

Thulin, L., "The First Personality Test was Developed during World War I", 23 September 2019, *The Smithsonian Magazine*, https://www.smithsonianmag.com/, accessed 6 October 2024.

Urla, J. and Jennifer Terry, J., "Introduction: Mapping Embodied Deviance", *Deviant Bodies*, Jennifer Terry and Jacqueline Urla, eds, Bloomington and Indianapolis, Indiana University Press, 1995.

Van der Waag, I., *A Military History of Modern South Africa*, Johannesburg and Cape Town, Jonathan Ball Publishers, 2015.

Verster, J.M., "Simon Biesheuvel, 1908–1991", *South African Journal of Psychology*, Vol 21, Issue 4, 267–270, 1991.

Ward, C., "Brief History", 10 December 2015, *CSIROpedia*, https://csiropedia.csiro.au/our-history/, accessed 24 April 2024.

"Sir Robert Watson-Watt, Engineer, Inventor, And Pioneer Of RADAR", https://engineeringhalloffame.org/profile/robert-alexander-watson-watt, accessed 2 March 2023.

Weinerlein, A.V., "War Reminiscences: War experiences of a youngster in South Africa, 1939–1945", *Military History Journal*, Vol 2, No 2, 2001, http://samilitaryhistory.org/vol122aw.html, accessed 8 June 2024.

Wessels, A., "South Africa and World War II: the decisive first two years on the home front (September 1939 - September 1941)", *Southern Journal for Contemporary History,* Vol 24, No 2, 1–24, 1999.

Wilkins, D., "RAF Training in Southern Africa", 25 July 2003, *WW2 People's War*, https://www.bbc.co.uk/history/ww2peopleswar/stories/37/a1122337.shtml, accessed 1 May 2025.

Index

S. Chetty, *Science and Technology in South Africa, 1939-1946*, Palgrave Studies in the History of Science and Technology,
https://doi.org/10.1007/978-3-032-05084-7

F

Q

R

X

Y

Z

GPSR Compliance
The European Union's (EU) General Product Safety Regulation (GPSR) is a set of rules that requires consumer products to be safe and our obligations to ensure this.

If you have any concerns about our products, you can contact us on

ProductSafety@springernature.com

In case Publisher is established outside the EU, the EU authorized representative is:

Springer Nature Customer Service Center GmbH
Europaplatz 3
69115 Heidelberg, Germany

www.ingramcontent.com/pod-product-compliance
Lightning Source LLC
Chambersburg PA
CBHW061223091225
36481CB00004B/82
9783032050830